THE DANCE OF LIFE

Other books by Edward T. Hall

BEYOND CULTURE
THE HIDDEN DIMENSION
THE SILENT LANGUAGE
HANDBOOK FOR PROXEMIC RESEARCH

With Mildred Hall

THE FOURTH DIMENSION IN ARCHITECTURE:
The Impact of Building on Man's Behavior

The Dance of Life

The Other Dimension of Time

EDWARD T. HALL

ANCHOR PRESS/DOUBLEDAY
Garden City, New York   1983

Library of Congress Cataloging in Publication Data

Hall, Edward Twitchell, 1914–
    The dance of life.

    Bibliography: p.
    Includes index.
    1. Time. I. Title.
BD638.H275   1982   115
ISBN 0-385-15964-1
Library of Congress Catalog Card Number 81–43650

This book is dedicated to
Mildred Reed Hall

CONTENTS

FOREWORD

No book is ever completed without the active help and participation of a great many people, some of whom are known and readily identified because they are present, actively involved, and their contributions are so patently obvious that recognition is unavoidable. There is another group, however, which remains forever hidden. I refer to those on whose shoulders the author stands while formulating his thinking—his intellectual forebears and those pioneers and groundbreakers who laboriously advanced our understanding of human behavior for hundreds of years before. My first acknowledgment is, therefore, to those scholars, known and unknown, recognized and unrecognized, living and dead, who have contributed so much to my thinking, made many of my insights possible, and without whom my own formulations would be nothing.

There are of course specific individuals whom one can identify as well as the tasks they have performed. Yet, even here, the contribution is somehow minimized because, having selected several, one automatically thereby excludes others. A good editor does almost as much in helping to mold a book as the author. A chance remark from a friend or reader frequently provides a sought-after link between disparate trains of thought that the author has been trying to connect. I, for one, feel quite humble and deeply grateful for the help I have received from those around me who have made this book possible.

William Whitehead edited the original manuscript and was, in addition, the source of much needed advice and counsel when it was in the crucial formative stages. I am especially grateful to my Doubleday editor, Sally Arteseros, for her enthusiastic support, her patience, and her professional experience. My agent, Carl Brandt, has always played two roles, both of which are important to any author. For me, he has represented the public and in that capacity he provides an unbiased mind which responds creatively to what I have written. In addition, he has provided encouragement during difficult times—and there are always difficult times. My partner and wife, Mildred Reed Hall, to whom I have dedicated this book, has performed functions too numerous to mention, in conceptual, editorial, critical, supportive, and backstopping capacities. Susan Rundstrom typed several versions of the manuscript. Susan has been so helpful and useful that I do not know what I would have done without her. Pat D'Andrea read and criticized the manuscript as well as prepared the index—a task on which much of the future usefulness of any volume depends. My colleagues Lawrence Wylie and William Condon both provided counsel, encouragement, and intellectual stimulation. My colleague Barbara Tedlock was kind enough not only to lend me a prepublication version of her book *Time and the Highland Maya* but also to discuss this material in some depth, allowing me to take full advantage of her rather unusual perspective on Quiché culture. To these people and unnamed others I must express my deep appreciation and everlasting gratitude.

May 4, 1982
Santa Fe, New Mexico

INTRODUCTION

The subject of this book is time as culture, how time is consciously as well as unconsciously formulated, used, and patterned in different cultures. Because time is a core system of all cultures, and because culture plays such a prominent role in the understanding of time as a cultural system, it is virtually impossible to separate time from culture at some levels. This is particularly true of primary level culture, about which I will be saying more.

The Dance of Life is one of several books about human beings, culture, and behavior. It deals with the most personal of all experiences: how people are tied together and yet isolated from each other by invisible threads of rhythm and hidden walls of time. Time is treated as a language, as a primary organizer for all activities, a synthesizer and integrator, a way of handling priorities and categorizing experience, a feedback mechanism for how things are going, a measuring rod against which competence, effort, and achievement are judged as well as a special message system revealing how people really feel about each other and whether or not they can get along.

Time is a core system of cultural, social, and personal life. In fact, nothing occurs except in some kind of time frame. A complicating factor in intercultural relations is that each culture has its own time frames in which the patterns are unique. This means that to function effectively abroad it is just as necessary to learn the language of time as it is to learn the spoken language. Several chapters in this book deal with the Americans and the Japanese as mirror images of each other, in which the determining threads of time set the stage for everything else. Other chapters are devoted to relations between Western European countries, as well as among Latin American, Anglo American, and Native American peoples.

One of the themes of this book is that human beings live in a single world of communication but they divide that world into two parts: words and behavior (verbal and nonverbal). Words, representing perhaps 10 percent of the total, emphasize the unidirectional aspects of communication—advocacy, law, and adversarial relationships—while behavior, the other 90 percent, stresses feedback on how people are feeling, ways of avoiding confrontation, and the inherent logic that is the birthright of all people. Words are the medium of business, politicians, and our world leaders, all of whom in the final analysis deal in power, so that words become the instruments of power. The nonverbal, behavioral part of communication is the provenance of the common man and the core culture that guides his life. This complex of feedback, local wisdom, and feelings is generally ignored or disparaged by our leaders. The question is: How is it possible to maintain a stable world in the absence of the feedback from the other 90 percent of communication?

In the above concept it is necessary to say something about culture, about which there has been considerable misinformation and not a little folklore. There are those who think of culture as something promulgated by anthropologists. Culture is not just a concept invented by anthropologists, any more than stratigraphy is a concept invented by geologists or evolution by Darwin. Culture is no more a concept than earth, air, or water. All of these things—including evolution—exist completely independent of what people believe. There are, of course, conceptual aspects of culture—i.e., our belief systems concerning the nature of culture which are analogous to the belief systems concerning the universe. Simply believing in something, however, doesn't make it so, and indeed, if what is believed is quite wrong, any action based on these beliefs can lead to dissonance and worse.

In taking the position that time and culture are inseparable in certain circumstances, I find myself on the opposite side of a high fence from many Western social scientists who, like pre-Copernican philosophers, hold that Western philosophical scientific models and, by association, Newtonian models are applicable to all cultures. They see time as a constant in the analysis of culture, and they also see Western science and Western thought as more advanced than other systems of thought. This position is epitomized by Yale University's Leonard Doob,[1] who has written extensively on time in the cross-cultural context. Doob views time as an absolute, rejecting the seminal anthropological studies of Africanists E. E. Evans-Pritchard on the Nuer and Paul Bohannan on the Tiv with regards to time. Doob's contention is that the time system is unrelated "to other cultural developments." I hold the opposite opinion: that time has everything to do not only with how a culture develops, but also with how the people of that culture experience the world. The English anthropologist E. R. Leach,[2] holding still another view of time in relation to culture, says: ". . . we create time by creating intervals in life. Until we have done this, there is no time to be measured." Implicit in this approach is the old Newtonian view of time as an absolute. As we shall see in the course of this book, making time contingent on measurement only accounts for one or at the most two of the many kinds of time and eliminates from examination people like the Hopi and the Sioux, neither of whom even has a word for time in their vocabulary. Each has time, however. The Hopi sun priests make accurate observations of the solstices and maintain a calendar of religious ceremonies. It is not necessary to belabor this point, but to deal with time according to Leach's view not only results in an oversimplification but also eliminates some of the more interesting, as well as basic, considerations of time.

My goal in this book is to use time as a means of gaining insight into culture, but not the reverse. In fact, I am not sure that the latter is possible; or if it is possible, then it is so only in a narrow sense. This has rather deep implications for our view of culture, as well as for mankind in general. There is a basic point that must be introduced here, because most of what follows subsumes it; namely, there is an underlying, hidden level of culture that is highly patterned—a set of unspoken, implicit rules of behavior and thought that controls everything we do. This hidden cultural grammar defines the way in which people view the world, determines their values, and establishes the basic tempo and rhythms of life. Most of us are either totally unaware or else only peripherally aware of this. I call these hidden paradigms primary level culture. Primary level culture (PLC), core culture, or basic level culture (I have used all these terms) is somewhat analogous to the hardware of a computer. Conscious, explicit, manifest culture, the part that people talk about and can describe, is analogous to the software—the computer programs. The computer analogy is oversimplified, but it will suffice for the moment. Carrying the analogy a step further: most intercultural relations are conducted as though there are only slight differences in the software and none in the hardware, as though the only differences are those which are representative of explicit, manifest culture, while all of the underlying PLC are identical (i.e., "people are all the same underneath"). The results of treating members of other cultures as though we are all programmed in the same way can range from the humorous through the painful to the tragic and even destructive.

Primary level culture has core components which pattern our thinking and which give us sets of underlying assumptions for arriving at the "truth." This was brought home to me recently while discussing the Japanese with a friend, a brilliant man with an unusually fine mind. I realized that not only was I not getting through to him, but nothing of a substantive nature that I had said made sense to him. He was operating on one set of assumptions—which we shared but which he also had never questioned—and I was describing a culture based on an altogether different set of assumptions. For him to have understood me would have meant reorganizing his thinking. It was as though I had suddenly imposed a new language with an entirely different grammar. It would have meant, for the moment at least, his giving up his intellectual ballast, and few people are willing to risk such a radical move.

One of the principal characteristics of PL culture is that it is particularly resistant to manipulative attempts to change it from the outside. The rules may be violated or bent, but people are fully aware that something wrong has occurred. In the meantime, the rules remain intact and change according to an internal dynamic all their own. Unlike the law or religious or political dogma, these rules cannot be changed by fiat, nor can they be imposed on others against their will, because they are already internalized.

There are at least three different levels at which culture can be seen to function: (1) the conscious, technical level in which words and specific symbols play a prominent part; (2) the screened-off, private level, which is revealed to only a select few and denied to outsiders; and (3) the underlying, out-of-awareness, implicit level of primary culture (PL). Language plays a prominent part in the first two but is secondary in the third. This does not mean that PL culture is entirely nonverbal, only that the rules have not yet been formulated in words. As a consequence, many cultures that appear quite similar on the surface, frequently prove to be extraordinarily different on closer examination.

These underlying differences are what I set out to examine when I returned to the study of time after almost two decades devoted to proxemics (the study of people's use of space as a cultural artifact, organizing system, and as a communication system).

There have been many times in my life when luck and good fortune have been on my side, and the study of proxemics was one. If it hadn't been for years spent with my feet more or less firmly rooted in the unconventional but solid soil of the primary culture of space, I doubt I could have survived with my intellect intact while trying to make sense of the massive literature on time. Unlike the study of territoriality, where the British ornithologist N. E. Howard[3] opened up new vistas and avenues of approach, I found the world of time closing in on me. Of course, there was a vast and important body of data on biological clocks, but somehow it was different from the biological data on crowding. It didn't yield the same results that the ethologists' study of territoriality did. There were no mass deaths from people being pressured by time (or were there?). In addition, the biologists and ethologists who have done such an extraordinary job recording the spatial and territorial behavior of other life forms haven't come up with comparable material on time. If there ever was a body of work governed by words which epitomizes Western thinking, it is time. In fact, if one reviews the field not for insights into the nature of time, but as a giant case study of Western thought, then things begin to make sense.

Behind these highly articulate endeavors to define the nature of time there lies a firm but virtually unexamined foundation of assumptions accepted as reality that have been neither questioned nor tested. Many of these are simply artifacts of our own implicit, primary level culture.

Human beings have reached the point where they can ill afford the luxury of ignoring the reality of the many different cultural worlds in which humans live. Paradoxically, for the Westerner, the study of contrasting cultures can be an exercise in consciousness raising, which is one of the purposes of this book. As long as human beings and the societies they form continue to recognize only surface culture and avoid the underlying primary culture, nothing but unpredictable explosions and violence can result. My thesis is that one of the many paths to enlightenment is the discovery of ourselves, and this can be achieved whenever one truly knows others who are different.

Today's world is dominated by two great but completely different traditions and, if Robert Ornstein[4] and Tadanobu Tsunoda[5] are correct, each emphasizes different areas of the brain. I am not referring to Capitalism and Marxism, or to great political doctrines such as totalitarianism and democracy. I mean the linear, externalized logic that began with the Greek philosophers in the fifth century B.C. and culminated in Western philosophies and today's Western science, and, on the other hand, the inward-looking, highly disciplined Buddhist philosophies in which Zen plays a prominent part. Each has entered into a powerful transaction, molding man and, through man, nature. But each works in radically different ways. Nevertheless, the two traditions have much to gain and learn from each other.

Fritjof Capra's book *The Tao of Physics* is a courageous attempt to deal with this issue on the level of physics, philosophy, and mathematics, and it can be helpful to those who look to physical science for inspiration. However, my approach is somewhat different, and while I have great respect for the powerful theories of physical science and what they have taught mankind about the physical world, and for the many advances that science and technology have made, I am constrained to remind myself that life itself, and particularly life for the human species, is the ultimate value against which all else should be measured. Without people, technology means nothing. If the world's problems are to be solved, it will be by human beings, not by machines; the machines are only here to help us. Technology is an inevitable result of mankind's propensity to evolve outside his body. The record on this score is impressive, but it is now time for the human race to begin again to direct attention to human beings and the social institutions that make this technology possible. By focusing our attention outward, we have been diverted from the real task of life: the understanding and mastery of life itself. This is where our two great but very different philosophical traditions become increasingly relevant.

Our purpose should be to facilitate human interaction, to begin to turn ourselves around, and to loosen the unconscious grip of culture so that instead of being controlled by the past, human beings can face the future in quite a new and more adaptive way. In setting these objectives, I do not mean to give the impression that our task will be easy. On the contrary, it is probably more difficult than anything the human species has thus far attempted. Paradoxically, the individual steps to cultural and personal comprehension are not inevitably difficult. It is the changing of behavior, and the integration of new patterns that lead to greater self-knowledge, that tax us most. In this sense, the Zen masters are right.

PART I

Time as Culture

1 How Many Kinds of Time?

Some things are not easily bent to simple linear description. Time is one of them. There are serious misconceptions about time, the first of which is that time is singular. Time is not just an immutable constant, as Newton supposed, but a cluster of concepts, events, and rhythms covering an extremely wide range of phenomena. It is for this reason that classifying time, in the words of the English Africanist E. E. Evans-Pritchard,[1] "bristles with difficulties." At the microlevel of analysis one might say that there are as many different kinds of time as human beings on this earth, but we in the Western world view time as a single entity. This is incorrect, but it is the way we see it.

It is possible to philosophize endlessly on the "nature" of time. While such an exercise can be engaging and even at times enlightening, I have found it more productive to use a different approach. In my approach, behavior comes first and words follow. Looking at what people actually do (in contrast to what they write and say when theorizing) one quickly discovers a wide discrepancy between time as it is lived and time as it is considered. As people do quite different things (write books, play, schedule activities, travel, get hungry, sleep, dream, medi-

tate, and perform ceremonies), they unconsciously and sometimes consciously express and participate in different categories
of time. For example, there is sacred and profane time as well as
physical and metaphysical time. It is also quite clear that time
as Einstein defined it in the technical sense—the time of the
physicists—is not the same as engineering or technological time.
Engineers must be as precise as possible, but they do not under
ordinary circumstances have to take into account the fact that
Einstein's time is relative and depends upon the speed with
which the clock is moving in relationship to the speed of light.
Then there are also the biological clocks which one hears so
much about, which get out of phase when people travel by jet.
Anyone who has experienced jet lag has firsthand knowledge
of the conflict between two different time systems: biological
clock time and the time of the clock on the wall in a distant time
zone. If one is from the Mountain States or the West Coast, one
discovers with dismay that during the middle of the day in a
European capital, at the very moment when one is supposed to
be alert for a meeting or conference, one is overwhelmed by
fatigue. According to its built-in rhythms, the body has been
up all night, and it is now six or seven in the morning! Regardless of the activity or the clock on the wall in the new time zone,
the body is screaming, "It's time to go to bed and rest."

In other words, it can be demonstrated that, on the level of
core culture as well as surface manifest culture, most of us who
live in the industrialized world are using and distinguishing
between six to eight (of the nine) kinds of time that it is possible
to identify. What we have here is the basis of a folk taxonomy.
Folk taxonomies sometimes have more to recommend them than
one might suppose and are certainly more in tune with the way
people think and behave at the implicit (primary) level than
the classification systems promulgated by philosophers and social
scientists. There are sacred, profane, metaphysical, physical, biological, and clock times, but we have very little idea of how
they all fit together or how each affects our lives. In addition,
there are also at least two categories of time of which AE people
(that is, American and European people) are only marginally
aware. All of us, for example, are tied together in an endless
web of rhythms—rhythms that influence how parents relate to

their children as well as how people relate to each other on the job and in the home. In addition to rhythms, there are larger cultural patterns, some of which are even antithetical to each other and which, like oil and water, simply do not mix.

How does one proceed to classify these different kinds of time and do it in a rational way so that the interrelationships can be seen as a coherent system? To assist in the task of symbolically integrating the different time systems, I considered a mandala. A mandala is one of mankind's oldest classification devices; it is usually in the shape of a circle or a square and is comparable to a matrix in mathematics. The basic purpose is to show the relationship of various ideas to each other in a comprehensive, non-linear fashion.

Mandalas are particularly useful when one is dealing with paradoxical relationships, dissimilar pairs or clusters of activities which one's intuition indicates are related but which have not been previously associated, linked, or combined into a com-prehensive system. After I tried various combinations it became clear that the mandala was the most promising approach. It is important to arrive at the right combinations because the mandala should conform as closely as possible to the actual rela-tionships encountered in life. My mandala gradually evolved and now comprises the four complementary pairs as shown in the diagram on page 16.

Before we proceed, however, a few comments should be made about symbolic representation in general, and this mandala in particular. Symbols should always be viewed as tools and con-sciously distinguished from the events which they symbolize. Words and mathematical symbols are classic examples of how such tools can be manipulated in ways real events cannot.

In Albert Einstein's terms, time is simply what a clock says and the clock can be anything—the drift of a continent, one's stomach at noon, a chronometer, a calendar of religious cere-monies, or a schedule of instruction or production. The clock one is using focuses on different relationships in our personal lives. Each division in the mandala represents a radically different kind of clock. Viewed in this light and taking into consideration the different classes of time, it is important to note that the rules for understanding one category (one kind of clock) are not applicable to another category. It is hopeless to try to make sense of physical (scientific) time in terms of its opposite, metaphysical time, and vice versa, or to apply the rules of sacred time to profane time. These classes of time are like different universes with different laws. The mandala expresses their different natures and the relationships among them.

In discussing the different types of time, I will try to provide enough description so that the average reader can grasp what is included under each heading and get a general idea of the

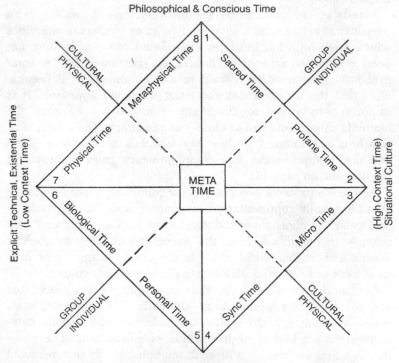

A MAP OF TIME

Philosophical & Conscious Time

Unconscious Emergent Time

Note: To discuss complementary systems it is necessary to invoke Meta Time, which is where the integrative concepts are located.

kind of a clock we are contemplating. The structural relationships between these various clocks and how they can be integrated are discussed in Appendix I.

Biological Time

Before life appeared on this earth—an estimated 2 to 4 billion years ago—the light and dark cycles, occasioned by the rotation of our tiny planet as it exposed first one and then the other side to the sun, represented but one of a series of cycles that made up an important part of the environment in which life evolved. The ebb and flow of the tides and the seasonal rhythms established by the travel of the earth in orbit around the sun formed the basis for other sets of clocks as life began. Sunspot cycles and the swelling and shrinking of the earth's primitive atmosphere like a huge beast breathing in its sleep all established a rhythmic change of environment that early life forms not only adapted to but eventually internalized.

From that point on, no form of life evolved or could evolve in a timeless nonrhythmic world. In fact, it was these very rhythmic changes from light to dark, from hot to cold, and from wet to dry that forced upon early living forms the very qualities that set the stage for later forms of life. Without environmental change, complex forms of life cannot evolve.

From this, one might say that in the beginning there was time and all time was periodic and rhythmic. As life evolved, the external cycles became internalized and took on lives of their own. Fiddler crabs, tuned to the twelve-hour tidal clock, foraged on beaches exposed at low tide. Oysters, tuned to the same clock, fed when covered with water. Grunions spawned and laid their eggs within thirty minutes of high tide. Even at the primitive level of slime molds, as many as six different times were delineated which control different states of developmental sequencing. Much higher up the evolutionary ladder, chickens lay more eggs when the days are longer in the summer. In humans there are hourly shifts in the hormonal levels of the blood.

Without intervention from the outside, these biological clocks will ordinarily stay in sync with the normal rhythms and cycles of the external environment. What happens inside is congruent with the outside world, so that while there are two kinds of time mechanisms, physical and biological, they behave as one.

In recent years there has been an enormous amount of research about how different organisms temporally integrate activities within as well as outside their bodies: sleeping, eating, mating, foraging, hunting, playing, learning, being born, and even dying, all involve exosomatic timing (phasing of an activity with events taking place outside the body). Phasing of this sort depends on an incredible system of internal timing mechanisms that keeps all living things in step not only with each other but also with the world. So important is this subject that over a thousand scientific papers a year are devoted to it.

Staying in phase is often taken for granted because it is an integral part of everyday life and much of it is out of our conscious control. Nothing can grow in a healthy way unless it is in a time-controlled, uniform manner; for example, it is the unregulated (out-of-phase) growth of cells in the body that characterizes cancer. Men have spent weeks in caves to see if their body rhythms are tied to the rising and setting of the sun or whether they are independent, trying to discover which of the two theories of biological timing actually determines how the "clocks" are set. Internal timing mechanisms are at work in the young man who hasn't discovered girls yet, who wakes up one morning feeling a new stirring in his blood and is amazed at how pretty the girl next door has suddenly become. He is responding to timing mechanisms deep within the glandular system in his body. Such mechanisms were laid down millions of years ago when the earliest mammals became differentiated from reptiles.

Anyone who has traveled east to west or west to east for more than three or four hours on a jet airplane and who has suffered jet lag has had firsthand experience in how our body rhythms are set according to the twenty-four-hour cycle of the planet. There is some speculation that travelers in space may run into serious difficulty not only because of lack of gravity, but also because of disorientation of the hundreds of biorhythms regulating body function. Professor Frank A. Brown of Northwestern

University postulates that physiological chaos will set in when humans travel too far beyond the boundaries of our own planet.[2]

The study of biological clocks includes not only those of living organisms, but also those of fossils—dating back to the Devonian times (350 to 400 million years ago) when the year was 400 days long rather than 365. On the behavioral side, the Japanese have been experimenting with biorhythms and keeping track of the periodicity of highs and lows in human energy, intellectual activity, and sociability. They report a reduced accident rate when their bus drivers drive more carefully during a "critical" phase. Whether the reduced number of accidents has come about as a result of simply telling the drivers to be careful at certain times, or because the rhythms have been accurately identified, has not been tested. However, biorhythms are closely related to "personal time," since they are supposedly unique to the individual.

Personal Time

Personal time has as its primary focus the experience of time (see chapter 8). Psychologists who have studied the way in which people experience the flow of time in different contexts, settings, and emotional and psychological states are concentrating their attention on personal time. Is there anyone who has not had the experience of time "crawling" or "flying"? Although biological time is relatively fixed and regular and personal time more subjective, there do seem to be environmental and physiological factors which help explain these great shifts in the way in which time is experienced. The slowing down of brain waves and the heart and respiratory rate during meditation have produced instances where people reported that "time stood still."

Physical Time

Sitting enclosed all day in "weatherized" steel and glass cocoons with all of our physical needs cared for, there is simply no reason for the average individual to keep close tabs on the sun in its annual pilgrimage from south to north and then back to south again. Specialists do that for us. June 21 and December 22 slip by unnoticed.

All over the world the pre-industrial peoples in the middle latitudes watched the sun travel along the horizon and carefully recorded its progress until the points farthest north (the longest day for those in the northern hemisphere) and farthest south (the shortest day for those in the northern hemisphere) were precisely charted and landmarks associated with the fixed sighting had been firmly established. In this way the date of all important ceremonies as well as planting and harvesting could be calculated for the coming six-month period. There are literally hundreds of such sighting points discovered in the American Southwest alone.

After observing that the sun moved and that everything was associated with that movement, early man had the problem of pinning down the pattern—of recording and fixing it in space in order to start counting the days. Whether the discovery was made once and diffused or was discovered independently all over the globe we will never know. One thing seems evident, however, and that is the practice of making accurate observations was tied to establishing the solstices. Ruth Benedict in her *Zuñi Mythology* describes this process: "The sun told him . . . come to the edge of the town every morning and pray to me . . . at the end of the year when I come to the south, watch me closely; and in the middle of the year in the same month when I reach the farthest point on the right hand watch me closely . . . The first year . . . he watched the sun closely, but his calculations were early by thirteen days. Next year he was early by twenty days . . . The next year his calculations were two days late. In eight years he was able to time the turning of the sun exactly."[3]

Observations such as these conducted in hundreds if not thousands of spots around the globe must have laid the foundations of modern science and have provided mankind with the first tangible clues that there was order in the universe, because these observations could be conducted hundreds of times and still yield uniform results.

Some of the greatest minds on this planet have focused their attention on physical time. Isaac Newton treated time as an absolute—one of the basic absolutes of the universe. Newton and his followers conceived of time as fixed and immutable, which meant that time could be used as a standard for measuring events. Newton was wrong, of course, as was clearly shown by Albert Einstein. Writing from his desk as a patent clerk in Bern, Switzerland, Professor Einstein provided compelling arguments that time was relative. He predicted that as a clock approached the speed of light, it would slow down. Einstein argued that a fast-moving astronaut could leave this earth and return a century later to find everyone he knew had died, while he himself had aged only a few years. This is not just a theory, but a physical fact, with far-reaching implications for mankind.

Fred Hoyle and other astronomers, who studied the shift toward red in the spectra of receding galaxies, date the universe around 15 billion years old, while the most distant objects from the solar system are 9 billion light-years away. To understand the meaning of these numbers of years or to reduce them to human scale is virtually impossible. There is nothing in human experience for the average person to compare them to. In the meantime, Newton's absolute time simply moved over into the profane box. No engineer could get along today without Newtonian time, which is an example of the prohibition against applying the rules of one time system to another. At the other end of the physical scale, there are short periods of time which, while important in the measurement of the drift of continents and radio waves, have little existential meaning to the man on the street. Such a clock at Johns Hopkins University's applied physics lab in Laurel, Maryland, measures time down to trillionths of a second—or about the time it would take a ray of light traveling at the rate of 186,000 miles per second to go the distance measured by the thickness of a playing card.

As is the case with the electromagnetic spectrum, the human sensorium is only able to respond perceptually to a small fraction of the visual part of that spectrum. The same applies to time. Mankind's capacity to experience time frequencies is limited to a tiny fraction of the total our instruments will tell us are present in the universe. Clearly the human species' time on earth is so short as to make it difficult to imagine when compared to the total.

Possibly even more remarkable than the effect of the 15-

billion-year reach of time on human consciousness is the fact
that physicists now see nothing sacrosanct about the directional-
ity of time. Admittedly, in the physical world on which we
depend for our intellectual underpinnings time is still irrevers-
ible, and H. G. Wells's time machines, while wonderful to
speculate about, are real only in the imagination. Nevertheless,
at the subatomic level, time may not be limited to one direction
only, but can go in both directions, forward and backward.
There isn't a clue as to what this will mean in the future—
reversal of aging, for example?

Metaphysical Time

There is no generally accepted theory of physical time to
account for metaphysical time. The two, metaphysical and
physical, make an interesting pair, and while Newton was
reported to have been more than a casual believer in the occult,
very few twentieth-century scientists will admit to preoccupations of this sort.

Leaving physics for the provenance of culture and the everyday life of human beings, one finds the metaphysical not only
alive and well but also thriving. For those who have experienced
it, the metaphysical has always been intimate and personal.
And though there is, so far, no way of equating the physical with
the metaphysical, this fact cannot be a basis for writing off the
mass of human experience, in all cultures, with this extraordinary
dimension. It is necessary to remember that one should not
attempt to apply the rules or methods of investigation of one
kind of time system to another, even a closely related one. We
must content ourselves with the view that they are simply
different, in the sense that words and things are different.

One of the foremost students of metaphysical time was J. B.
Priestley, who drew upon the television audiences of Great
Britain for hundreds of examples of individual experiences in
which individuals had transcended both time and space.[4] One
does not have to read Priestley to be aware of time warps. Most
people have had a brush with "déjà vu," for which I have no
explanation. However, as many have observed, a goodly number
of experiences of this sort, including those in Priestley's sample,

are not only trivial in nature but also characterized by an almost total lack of surrounding context. They are difficult, if not impossible, to pin down in either time or space. But not all of Priestley's examples are so easily dismissed. He seems to provide us with a fairly representative sample, as these things go— representative for the British Isles, that is. However, a significant number of the cases sent to him by his public are not at all trivial. Nor can they be brushed aside as the product of a demented mind. Those that are published—and many were not —are well documented. Also, in these instances the context was sufficiently explicit and vivid in detail so that Priestley's sources were actually able to act on their precognitive experiences to avoid catastrophes. We can't explain these things and right now I wouldn't even attempt to do so. The fact is that all over the world, regardless of culture or station in life, human beings keep reporting these rather extraordinary occurrences. I for one would be opposed to dismissing them simply because they don't happen to fit our particular paradigms. The metaphysical plays an important role in many people's lives, and it is comforting for them to know that it is there.

In this book, while little more will be said about the metaphysical, as a student and observer of human behavior I am compelled to take the position that, until proved otherwise in irrevocable terms, the metaphysical must be viewed as simply one more variant of what the human species has experienced which must be taken as seriously as anything else that happens to human beings. There are, however, a great many things about our own species which do not come under the metaphysical umbrella which are just as remarkable and much less well known. Some will be discussed in this book. In one sense I think it may be quite stubborn and blind on our part to treat the metaphysical as separate from "life."

Micro Time

Only recently identified and still not widely recognized, micro time is that system of time that is congruent with and a product of primary level culture. Its rules are almost entirely outside conscious awareness. It is culture specific; that is, it is unique to each culture. Monochronic and polychronic times (see chapter 3) are examples of major patterns of this type. Even though monochronic time is shared by most North European countries and cultures, it has its own variations for each culture and region. The significance attached to different periods of waiting time by Americans when they are on business is another example. Micro time is one of the basic building blocks of culture. Much of the material in this book is devoted to micro time.

Sync Time

Sync time is an even more recent discovery than its partner, micro time. The term "to be in sync" is derived from the media and dates back to the beginning of "talking pictures" when it was necessary to synchronize the sound track with the visual record on film. Since then, frame-by-frame analysis of motion picture film taken during normal transactions of daily life reveal that when people interact they synchronize their motions in a truly remarkable way. One of the first things that happens in life is for newborn infants to synchronize their movements to the human voice. People who are out of sync with a group are disruptive and do not fit in. Different people move to different beats. Each city and town in the United States has its own beat. Each culture has its own beat. Though it took the white man thousands of years to discover "sync time," the Mescalero Apaches have known its significance for centuries. Chapters 9 and 10 are devoted to sync time.

Sacred Time

Modern AE people—peoples of American-European heritage —have some difficulty understanding sacred or mythic time, because this type of time is imaginary—one is *in* the time. It is repeatable and reversible, and it does not change. In mythic time people do not age, for they are magic. This kind of time is like a story; it is not supposed to be like ordinary clock time and everyone knows that it isn't. The mistake is in trying to equate the two or to act as if it were necessary to create a fixed relationship between the sacred and the profane. When American Indian people participate in ceremonies, they are in the ceremony and in the ceremony's time. They cease to exist in ordinary time. For some, sacred time makes the rest bearable.

By putting themselves in sacred time, people subconsciously reaffirm and acknowledge their own divinity, but by raising consciousness they are acknowledging the divine in life. Mircea Eliade in his book *The Sacred and the Profane* sees this as imitating God (imitatio dei). I don't; I see it as defining consciousness.[5]

Profane Time

Rooted in the sacred time of the Middle East, which in turn grew out of physical time, profane time now dominates daily life and that part of life which is explicit, talked about, and formulated. In the Western world, profane time marks minutes and hours, the days of the week, months of the year, years, decades, centuries—the entire explicit, taken-for-granted system which our civilization has elaborated. Possibly because the time system is linked to the sacred in a complementary way, some of the sacredness rubs off and therefore people generally do not tolerate changes in it. When the Julian calendar had slipped so far out of line because of the computation of leap year, it was necessary to recalibrate it. Pope Gregory XIII tried to alter the calendar by cutting out ten days. People responded by rioting, shouting, "Give us back our ten days." There are other more contemporary examples: When President Franklin D. Roosevelt tried to change Thanksgiving so that it wouldn't be so close to Christmas, the public response was almost as intense as in Gregory's time.

Meta Time

Meta time is made up of all those things that philosophers, anthropologists, psychologists, and others have said and written about time: the innumerable theories, discussions, and preoccupations concerning the nature of time. It is not time in the true sense but an abstraction from different temporal events. Much of the confusion or lack of consistency between the many theories of time are due to different individuals looking at one kind of time (metaphysical time, for example) from the perspective of another (physical time) or confusing meta time with reality.

2 Different Streams

> *Truth only reveals itself when one gives up all preconceived ideas.* SHOSEKI[1]

When visiting the Hopi villages for the first time in 1931, I didn't need H. G. Wells and his time machine to move into another age; for there it was, floating like a mirage on the mesas above. Soon I would be immersed in it. Even close to the mesas it was difficult to distinguish the houses from the weathered sandstone of the crumbling vertical cliffs. Those mesas and arroyos, those distances and roads set that world apart, preserving a life-style and culture out of the past. Here and there were islands of the white man's world (like icebergs in an Arctic sea), but unlike the icebergs those islands and their impact would grow and infiltrate the Indian's world.

The country posed problems for the tourists then, many of whom would have mini tantrums when time stopped, as it was frequently apt to do. Standing red-faced on the side of a running arroyo, they champed at the bit because they couldn't get across. Their automobiles were immobilized, and they would have to wait for the water to go down (which could take anywhere from five to thirty-five hours). I never realized that bridges were an agent of time until I worked on the reservation. Now, of course, everything on the mesa is an agent of time. The

white man's world has taken over and there are not only paved
roads, but also bridges all the way across the reservation. A wonderful scenic trip that put one in the middle of the country, to
say nothing of into another age, and used to take a week is now
completed in a matter of hours.

Having visited the Hopi and Navajo reservations in 1931, I
never dreamed that I would be back within a year, and would
spend significant portions of the next five years on those reservations. It started with a telegram from John Collier—Commis-
sioner of Indian Affairs—offering me an opportunity to participate in a new program. The objective was to do something
positive with and for the American Indian. As luck would have
it, I was sent to the Navajo-Hopi part of the reservation and
worked first as a manager running a camp for the two tribes
and then later as a construction foreman building dams and
fixing roads. It was in this context that I experienced for the
first time how cultures really do clash with each other and how
difficult it is to get behind the externals and down to the
nitty-gritty of what makes each group behave and think the
way they do. Having arrived at an insight, I learned how little
this means to the average white person on the reservation whose
job it is to work with my Native American friends.

I soon learned that I was dealing with at least four different
time systems: Hopi time, Navajo time, government bureaucratic
time, and the time used by the other white men (mostly Indian
traders) who lived on the reservation. There was also Eastern
tourist time, banker's time (when notes were due), and many
other variations of the white man's time system. And what
differences there were among those time systems! There seemed
to be no way at all to bring them in line. Even as a somewhat
naïve youth, I was amazed and puzzled by how little importance
was attached to those differences. They were ignored so that
everyone could adhere to his own time system.

To the Navajo, the future was uncertain as well as unreal,
and they were neither interested in nor motivated by "future"
rewards—a foundation on which many of our government programs were based. Sheep and stock reduction programs were
planned and sold to the Indians in terms of future rewards,
"when the range recovers from overgrazing" twenty years hence,

which the Navajo took as a ludicrous joke—simply one more example of white perfidy.

The government bureaucrat assumed erroneously that anything technical, such as a dam, a road, a building, or a boundary, was just that: a technical problem involving only a search for the correct technical solution. It did not come easily to us then or now that other issues might be at stake.

From the beginning, the government couldn't and didn't see it that way, and it managed to stumble into one mistake after another. You couldn't blame the engineers, reared in Oklahoma and recently reduced in force in Washington and sent out to the reservation on emergency short-term jobs. Those well-intentioned men knew nothing of either the Hopi or the country. Dams were staked out in drainages that looked promising, without consultation with the Hopi either for information on the micro-climate or on clan ownership. The Indians talked about how one drainage could be counted on for rain and runoff while another one—only three miles to the east—was always dry. They knew which watersheds would produce runoff. They were also fully aware of the social and political consequences of placing a dam in any place on that reservation.

Nor was the religious impact of the location projects considered. Would the dam be near clan-owned shrines and sacred places? In fact, the reservation superintendent—the man who, like the ship's captain, was in command and who set policy—once threatened to fire me and "run me off the reservation" because I suggested it might make sense to consult the Hopi. The result was that dams were frequently located so as to exacerbate ongoing feuds between clans and villages. A government dam built on clan-owned land for the use of all Hopi technically could be claimed by that particular clan; if a dam were made available to other clans for watering their stock, this could establish a precedent and provide an opening wedge for future claims on the land. These inauspicious beginnings couldn't help but influence the Hopi's attitude toward work on a dam. The fact that the two radically different tribes—Hopi and Navajo—inhabited the reservation didn't simplify matters, and there were further complications.

Two major differences between the Hopi and the Navajo, having a direct influence on the work, were immediately apparent. First, the Navajo preferred working a full month (a twenty-day work period without weekend breaks); and, there was an almost organic relationship between the Navajo men[2] and their work. They took great pride in good workmanship and identified deeply with each dam they built. Above all, they wanted the work done right. The Hopi, in those times, seldom had this sort of identification. When it came to work schedules, Hopi men, in contrast to the Navajo, were not only willing, but also preferred to divide up the work. Any period, no matter how short, was all right as long as it meant that more people from their village could be hired. Also, Hopi men had an addiction to devoting a portion of each day to cultivating their fields, something which failed to register with the dam builders. The men felt uncomfortable and unfulfilled when this ritual couldn't be performed. The government work interfered with their routines and kept them away from their fields. Actually, the government could have employed a completely new crew every day had it been possible to keep up with the bookkeeping. Computers weren't even a gleam, then, in John Von Neuman's eye. Everything was done by hand, and not very accomplished hands at that. Payrolls complicated by Indian names in a strange language, unfamiliarity with the Indians themselves, and the fact that many men had two or more names (white and Indian) posed more problems than the meager clerical staff could handle.

Because each dam required from one to three thousand mandays of labor to complete, it would have been theoretically possible to use every Hopi male on the reservation over the ninety-day period required to finish a given project. That no one would have had time to learn how to do his job properly was not considered by the Hopi. Nor did the fact that the completion of each project would have been endlessly delayed seem relevant to them. The whites simply took this back-and-forth discussion and wrangling as an additional sign of Hopi irascibility and another brick in the edifice of stereotypes about the Indian. It did not occur to most of the Indian Service personnel that they were dealing with a different mentality based on whole congeries of unfamiliar and exotic assumptions. That there could actually be different assumptions on matters of this sort which involved logic was simply unacceptable to them.

How Hopi time affected the whites in the area was more dramatic in its implications than the best science fiction. Many of the foundation stones on which our system was built were simply not present in Hopi culture. It gradually became apparent that the Hopi were operating on a very different plane.

AE people grow up expecting that, once initiated, a project will continue more or less without pauses or serious breaks until it has been completed. We Americans are driven to achieve what psychologists call "closure." Uncompleted tasks will not let go, they are somehow immoral, wasteful, and threatening to the integrity of our social fabric. A road that ends abruptly in the middle of nowhere signals that something really went wrong, "somebody goofed." I became increasingly puzzled as I began to realize that the Hopi lacked this crucial concern for closure and that they had no timetables in their heads for ordinary built objects, as contrasted with their ceremonies, which *were* scheduled. In those years, one of the most noticeable and arresting characteristics of Hopi villages was the proliferation of unfinished houses which dotted the landscape. Walls would be laid in courses of rock and mortar (beautiful stone walls—strong, solid, well made, showing that people cared and had invested considerable amounts of time and effort). The window frames—with or without windows—would be inserted, but the roof would not be finished. Vigas made from fir and pine from distant forests, cut and scraped, would be lying neatly stacked next to the house, waiting to be put in place to support the roof. Everything was ready—waiting for what looked like three weeks' work on the part of two or three men—and there the house would sit for years. Questions by whites as to when the house would be finished were treated as non sequiturs—which they were to the Hopi. There appeared to be no built-in time schedule, no feeling that life would be out of joint as long as the task remained incomplete. There was no apparent relationship between the completed project and a schedule for completion, all of which we whites took for granted. Yet the underlying

culture pattern that explained the unfinished houses had an unexpected impact on the conservation work—which I'll describe
in a moment.

In regard to Hopi demands for more frequent crew changes,
the government compromised, changing crews every two weeks.
This only meant double the paperwork, and double the number
of checks. The real trouble began, however, when it came time
to divide the number of man-days spent on the job (the labor
cost) by the cubic yardage of a completed dam. In those days,
comparable work on regular construction projects would run
from 60 to 75 cents a cubic yard of earth moved. According to
our engineers, the dams constructed by our Hopi crews were
costing from $4.50 to $5.00 a yard—six to ten times the normal.
It was evident that the Hopi weren't working as hard as they
should be—at least that was the implication. When this was
pointed out to the crew foremen, the Hopi response was strong
and negative. Deeply incensed by the white man's criticisms,
they bitterly complained of harassment. The program developed
into a logrolling boondoggle, not because the Hopi couldn't or
wouldn't work but because no one felt it was important to make
the effort to explain what the work was all about in terms understandable to the Hopi.

Still another side of this complex issue must be raised in explaining the Hopi behavior. Basically, the cultures of the world
can be divided into those in which time heals and those in
which it doesn't. Whites belong to the first category and the
Hopi belong to the second. For the Hopi, past experience with
whites—first the Spanish and then the white Americans from
the East in the late nineteenth and early twentieth century—is
as sad a story as one can find in the archives of colonialism. The
Spanish priests enslaved the Hopi. Our government didn't do
much better. The Indians were thought of as heathen savages
who must be turned into white men as expeditiously as possible.
Their sacred ceremonies were disrupted and even banned. Everything possible was done to destroy the fabric of Hopi life, but
it was the attacks on their religion that caused the bitterest
resentment. Religious leaders were jailed—taken away from their
families, who were not informed about what had happened to

their fathers and husbands or when, if ever, they would be returned to their homes, so that wives remarried, not knowing their husbands would return. In another instance, an entire village—men, women, and children—was treated as sheep—literally—and run through cement troughs full of Black Leaf 40, a strong concentration of pure nicotine. The excuse was that the people had lice!

This is not a story for anyone to be proud of, yet, because we live in a "time heals" culture, the white people of the Indian Agency at Keams Canyon were either completely ignorant of the past or else assumed that because "that was ancient history" the Hopi could not feel intensely about things which happened before those now living were born. Well, the Hopi hadn't forgotten, and for them the past dominated the present. Also, our government was oblivious to the life in the Hopi villages. Most of what transpired on the mesa tops simply did not exist in the minds of the whites. This was because employees didn't spend any more time than they had to in the villages. In fact, there was an informal rule that one should not get too involved with the Indians. (I was later to see this attitude in our overseas diplomatic and technical aid missions for Third World countries.) Separated spatially, temporally, and culturally, the Hopis fumed. Past injustices gnawed at them and wouldn't let up. The visions of injustices grew while the circumstances that had led to them were forgotten. Then along came the United States Government with its "Emergency Conservation Work" program and hit them with some ridiculous nonsense (to the Hopi) about how many days it is supposed to take to finish a dam, as though the dam had a built-in schedule like the maturing of a sheep or the ripening of corn. It was just one more instance of the white man making life difficult. The dam was no different, in the minds of the Hopi, than their unfinished houses.

The tragedy was that, with the best of intentions, the government was really trying to do something for the Hopi: to give them work in a time of depression, work designed to improve their country. Money was cranked into the economy, dams and roads were built, springs were developed, but the end result was that old rivalries were revived because of the construction

schedules and the placement of dams that ignored traditional
boundaries, which the government didn't even know existed. A
knowledge of primary level culture simply was not ours.

These new "injustices" were perpetrated because the government, in its naïveté, thought that the Hopi would naturally want
to get as much value from the government appropriation as
possible. It was up to them; they could build twenty dams or
two. The alloted funds were the same. It soon became apparent
that we were at odds because of two systems of logic which
were diametrically opposed. Furthermore, there was no readily
identifiable common ground on which to meet.

The problems we were having all had either a strong spatial
or temporal component or both. Who was there that I could talk
to about this? With the exception of one Indian trader (a friend
named Lorenzo Hubbell) who taught me most of what I knew
at the time about the Hopi and the Navajo, no one seemed to
have a clue as to what it was that was causing the difficulty or,
in many instances, that there even *was* any difficulty. What they
were experiencing was considered normal for work with American Indians. Trying to explain why it was that the Hopi were
reacting as they did, I found that most people either were not
interested or else chalked it all up to irascibility, saying, "The
Hopi are always complaining." The tragedy is that the same
sort of misunderstanding exists today. In fact, in spite of modern-
ization on many fronts it's almost as though disaffection between
the two cultures has increased instead of diminished. The cultural gap today is as broad and as deep as it ever was. Admittedly, there are many built-in obstacles in the road to understanding, and differences in the structure of the two languages
is one of them.

Let's consider the way the Hopi language influences the way
they think. What follows is based directly on the work of a
great pioneer thinker in the field of linguistics, Benjamin Lee
Whorf, a linguist and chemical engineer. Whorf's theory is not
only technical but also detailed.[3]

All AE languages, including English, treat time as a continuum divided into past, present, and future. Somehow we have
managed to objectify or externalize our imagery of the passage
of time, which makes it possible for us to feel that we can

manage time, control it, spend it, save it, or waste it. We have a feeling that the process of "becoming later" is real and tangible because we can attach a numerical value to it. The Hopi language does *not* do this. No past, present, or future exists as verb tenses in their language. Hopi verbs have no tenses, but indicate instead the validity of a statement—the nature of the relationship between the speaker and his knowledge or experience of that about which he is speaking. When a Hopi says, "It rained last night," the hearer knows how that Hopi speaker knew it rained: whether he was out in the rain and got wet, looked outside and saw it raining, whether someone came through the door and said it was raining, or he woke up in the morning and saw that the ground was wet and assumed that it had rained.

In AE languages, temporal terms such as summer and winter are nouns, which gives them a material quality because they can be treated like any other noun, numbered and given plurals. In other words, they are treated as objects. The Hopi seasons are treated more like adverbs (the closest AE analog). The Hopi cannot talk about summer being hot, because summer is the quality hot, just as an apple has the quality red. Summer and hot are the same! Summer is a *condition:* hot. There is nothing about summer that suggests it involves time—getting later—in the sense that is conveyed by AE languages.

It is clear that our emphasis on saving time, which goes with quantifying time and treating it as a noun, would also lead to a high valuation of speed, which is demonstrated in much of our behavior.

Living in the eternal present as the Hopi do and spending the "now" preparing for ceremonies, one feels that time is not a harsh taskmaster nor is it equated with money and progress as it is with AE peoples. For AE peoples, it does have that characteristic of adding up, of never letting them forget. This can be burdensome. To the Hopi, the experience of time must be more natural—like breathing, a rhythmic part of life. Also, the Hopi, to my knowledge, have never become preoccupied with philosophizing about the "experience" of time, or the nature of time.

Philosophy is one of the cornerstones on which the intellectual edifice of America is built. Science is another, and technology makes up a third. Religion, however, is not as central as it once was, and it has for some time now been separated from daily life, sealed away in a compartment of its own.

Not so for the Hopi.

Religion is the central core of Hopi life. Religious ceremonies perform many functions which in AE cultures are treated as separate and distinct entities, quite apart from the sacred: disciplining children, for example; encouraging rain and fertility; staying in sync with nature; helping the life-giving crops to be fertile and to grow; relating to each other; and initiating the young into adulthood. In fact, religion is at the center not only of social organization but also of government, which is part and parcel of Hopi ceremonial life.

The Hopi year is divided into two halves separated by the solstices. The Kachinas, masked figures,[4] are somewhat analogous to gods or nature spirits or even the embodiment of dominant themes in Hopi life. They live with the people for half of the year and return to their home in the San Francisco Mountains (north of Flagstaff, Arizona) for the remaining six months. Every man, woman, and child is initiated into the Kachina cult and participates in the Kachina ceremonies. The year begins with the rites of the winter solstice, at which time everything is prepared for the coming year. The fixing of the exact date of the winter solstice is extremely important. It is the Sun Priest's duty to determine precisely when the sun has stopped on its journey to the south and, going no farther, is turning around and about to move north again.

Living with the Hopi, talking to them about the dances, watching the dances being performed, I found myself many times enveloped in a particular kind of time and space warp which only occurs if the dance is successful. When this happens all consciousness of external reality, all awareness of the universe outside, is obliterated. The world collapses and is contained in this one event; there is nothing else, nothing except the people, the crowded kiva, and the dancers.[5] If this could happen to me, a young white, think how it must seem to the Hopi!

Hopi time, when they talk and now write about it today, is most frequently in terms of the dance ceremonies held throughout the year. "It was just before Wowochim." "It happened during Soyal." Wowochim is the tribal initiation ceremony held in November. Soyal follows Wowochim in December and is associated with the winter solstice. It is an important ceremony, participated in by the Sun Clan in honor of the sun god (the most powerful and important god, who controls all life).[6] The ceremony celebrates or marks his departure from his southern home and the beginning of his journey north to his northern home, where he spends the summer.

Associating events with sacred dances quietly reinforces the power and strength of these events. Life, in fact, revolves around them, and for the members of the clans and secret societies who perform the dances, they take precedence over everything else: work, family obligations, sex, and personal feelings and commitments. Nothing was supposed to be more important, and for those who were initiated nothing was, particularly in the old days. Another difference between AE time and Hopi time is that in the AE pattern the public ceremony is where it all comes together, whereas in the Hopi tradition the public part of the ceremony is not only preceded by days of preparation in the kiva but also followed by several more days of kiva rituals.

The Hopi marriage ceremony illustrates another variant of the culture pattern. In the AE cultures the marriage marks a clearcut division between one state and another. Who hasn't heard a young woman who has just exchanged marriage vows comment: "I can't get over it. I am now Mrs. Walter A. Nash. Just a few minutes ago I was plain Jane Moore and single and now I am a married woman!" That such remarks are so common reflects the fact that even in AE cultures people find it surprising that such incredibly important changes can take place in a matter of minutes. With this buildup, one would expect that the Hopi marriage would take somewhat longer to solemnize, which is precisely the case. Second Mesa artist Fred Kabotie's and Sun Chief Don Talayesva's descriptions of Hopi marriages[7] included as many as twenty-six different events spanning an entire year. The Hopi don't just slip into something overnight. It is important to stress the point that differences of this sort are not just conventions, but reflect deep-set structural differences between the two cultures. In AE cultures we expect that things will happen fast once we have made up our minds and, as a consequence,

we are apt to pay little attention to the pattern we are weaving
in life's fabric or to the slow accumulation of Karma in the
multiple acts of daily living.

Living and working on the Hopi and Navajo reservations in
the early '30s was an extraordinary experience. There were two
relatively intact exotic societies, each different from the other,
and both of them different from our own. In addition, there was
the environment, and the physical setting, a life-style fifty years
behind that of the rest of the country. Summer's rains and winter's snow and mud coupled with the lack of roads created a
situation with its own rules. We didn't make the rules, the
country and seasons made the rules. When it snowed or rained,
everything stopped (except the horses and wagons, of course,
which were adapted to mud). When the washes ran, no one in
his right mind would attempt to cross—even the small ones. The
rusted carcasses of too many automobiles and trucks lay buried
in the sandy floors of the Dinnebito, Oraibi, Polacca, and Weepo
washes. The general practice—even for whites—when confronted with a running arroyo was simply to wait until the
water went down. One would cross the sandy floor of those
arroyos which were up to a quarter mile across and forty feet
deep only when they were dry. Those stream beds would be
dry for months at a time, and then one day, out of the north,
there would be a distant roar that became louder by the minute
as a wall of water pushing brush and logs and trees in front of
it would round a bend. What had been dry sand was transformed
into a raging brown torrent in a matter of minutes. Arroyos
would run for hours and even days with the runoff from a
cloudburst up to fifty miles upstream. It helped to know about
these things—the fact that events many miles away could result
in the loss of a good vehicle stuck in what had been dry sand
only minutes before. One learned to pay attention to what was
happening at all points of the compass, but particularly upstream, because all the main washes ran roughly from northeast
to southwest. Thunderclouds over Black Mesa could mean floods
in Oraibi five hours later. Those things, like many others, were
important because one frequently could be left high but not
dry waiting on the wrong bank of an arroyo for days until
roads were passable again. I soon learned that it did not pay

to resist the country but rather to move with it, becoming an integral part of it, which is difficult for a white man.

This message came across to me in another unexpected way because of my horses. I had always had horses. They went with the schools I attended and had become an integral part of my life. The only thing was, my horses were in Santa Fe, New Mexico, and I wanted to move them to the reservation. Trucking them out was impractical and too expensive, so I decided to ride them. The experience proved to be one which left an indelible mark on my psyche. Saddling up two horses and using a third as a packhorse, two of us rode from Santa Fe, over the Jemez Mountains, across the Continental Divide where it intersects Chaco Canyon, past Crown Point and Gallup, over to Window Rock, through the cool forests of the Fort Defiance Plateau, dropping down to Ganado and from there striking due west across the open flats and sagebrush-covered mesas and on to Jeddito, Keams Canyon, Polacca, Toreva, Chimopovi to Oraibi and, after a pause to see Lorenzo Hubbell, north to Pinyon, a sort of epicenter of Navajo spiritual life.

Our daily average was twelve to fifteen miles, otherwise the mustangs we were riding would tire and ultimately give out. Dropping down from the fir-covered slopes of the Jemez Mountains onto the parched plains to the west, I watched the same mountain from different angles during three days, as it seemed to slowly rotate while we passed by. Experiences of this sort give one a very different feeling than speeding by on a paved highway in one or two hours. The horse, the country, and the weather set the pace; we were in the grip of nature, with little control over the rate of progress.

Later, riding horseback on a trek of three or four hundred miles, I discovered it took a minimum of three days to adjust to the tempo and the more leisurely rhythm of the horse's walking gait. Then I became part of the country again and my whole psyche changed. I used to notice that the cowboys, some of whom helped raise me when I was a preadolescent, had a tempo of speech unlike that of other white men. It didn't speed up and slow down to keep in sync with the people around them. As a group, they were marked by their own tempo—geared to the personality, the mood, and the situation. During emergencies—such as when the horse and the pack animal I was leading fell off a mountain trail and landed on a ledge—the words came fast as lightning: "For Chrissake, get that #°x#@! rope off the pummel or you'll get cut in half if the horse gets up." But for conversation they did not want to be rushed, so they set their own pace. Dudes and tourists would seek quick responses to their questions and verbally tailgated these men of the outdoors, trying to "get them up to speed." They never realized that it was their own urban tempo that was out of sync with the body and that the mere rush of words over the years could erode the disposition just as surely as those Western arroyos eroded the soft soil of the valley flats through which they ran.

Cultural reaction time—the time required for a response to a threat, challenge, slight, or an injustice—varies greatly. Like booby traps and land mines set to go off, the time interval can be anything from days to years for major events. White Americans, when stimulated to action, tend to respond somewhat precipitously, whereas many Native Americans burn with a much longer fuse. It takes them a while to build up to action. For example, the Pueblo Revolt of 1680 drove the Spaniards from the Southwest. It wasn't until 1692 that the Spanish returned. Some of the Indians allowed the priests to set up their churches again, some did not. Awatovi, the easternmost of the Hopi villages, where the Spanish had established a thriving mission, took the priests back. It was six years before the other Hopi acted in revenge. They destroyed Awatovi, murdering the inhabitants.

Whites tend to think that because nothing overt is happening, nothing is going on. With many cultures there are long periods during which people are making up their minds or waiting for a consensus to be achieved. We would do well to pay more attention to these things.

3 Monochronic and Polychronic Time

Lorenzo Hubbell, trader to the Navajo and the Hopi, was three quarters Spanish and one quarter New Englander, but culturally he was Spanish to the core. Seeing him for the first time on government business transactions relating to my work in the 1930s, I felt embarrassed and a little shy because he didn't have a regular office where people could talk in private. Instead, there was a large corner room—part of his house adjoining the trading post—in which business took place. Business covered everything from visits with officials and friends, conferences with Indians who had come to see him, who also most often needed to borrow money or make sheep deals, as well as a hundred or more routine transactions with store clerks and Indians who had not come to see Lorenzo specifically but only to trade. There were long-distance telephone calls to his warehouse in Winslow, Arizona, with cattle buyers, and his brother, Roman, at Ganado, Arizona—all this and more (some of it quite personal), carried on in public, in front of our small world for all to see and hear. If you wanted to learn about the life of an Indian trader or the ins and outs of running a small trading empire (Lorenzo had a dozen posts scattered throughout north-

ern Arizona), all you had to do was to sit in Lorenzo's office for
a month or so and take note of what was going on. Eventually
all the different parts of the pattern would unfold before your
eyes, as eventually they did before mine, as I lived and worked
on that reservation over a five-year period.

I was prepared for the fact that the Indians do things differently from AE cultures because I had spent part of my childhood on the Upper Rio Grande River with the Pueblo Indians
as friends. Such differences were taken for granted. But this
public, everything-at-once, mélange way of conducting business
made an impression on me. There was no escaping it, here was
another world, but in this instance, although both Spanish and
Anglos had their roots firmly planted in European soil, each
handled time in radically different ways.

It didn't take long for me to accustom myself to Lorenzo's
business ambiance. There was so much going on that I could
hardly tear myself away. My own work schedule won out, of
course, but I did find that the Hubbell store had a pull like a
strong magnet, and I never missed an opportunity to visit with
Lorenzo. After driving through Oraibi, I would pull up next to
his store, park my pickup, and go through the side door to the
office. These visits were absolutely necessary because without
news of what was going on life could become precarious.
Lorenzo's desert "salon" was better than a newspaper, which,
incidentally, we lacked.

Having been initiated to Lorenzo's way of doing business, I
later began to notice similar mutual involvement in events
among the New Mexico Spanish. I also observed the same patterns in Latin America, as well as in the Arab world. Watching
my countrymen's reactions to this "many things at a time" system I noted how deeply it affected the channeling and flow of
information, the shape and form of the networks connecting
people, and a host of other important social and cultural features of the society. I realized that there was more to this culture
pattern than one might at first suppose.

Years of exposure to other cultures demonstrated that complex
societies organize time in at least two different ways: events
scheduled as separate items—one thing at a time—as in North

Europe, or following the Mediterranean model of involvement in several things at once. The two systems are logically and empirically quite distinct. Like oil and water, they don't mix. Each has its strengths as well as its weaknesses. I have termed doing many things at once: Polychronic, P-time. The North European system—doing one thing at a time—is Monochronic, M-time.[1] P-time stresses involvement of people and completion of transactions rather than adherence to preset schedules. Appointments are not taken as seriously and, as a consequence, are frequently broken. P-time is treated as less tangible than M-time. For polychronic people, time is seldom experienced as "wasted," and is apt to be considered a point rather than a ribbon or a road, but that point is often sacred. An Arab will say, "I will see you before one hour," or "I will see you after two days." What he means in the first instance is that it will not be longer than an hour before he sees you, and at least two days in the second instance. These commitments are taken quite seriously as long as one remains in the P-time pattern.

Once, in the early '60s, when I was in Patras, Greece, which is in the middle of the P-time belt, my own time system was thrown in my face under rather ridiculous but still amusing circumstances. An impatient Greek hotel clerk, anxious to get me and my ménage settled in some quarters which were far from first-class, was pushing me to make a commitment so he could continue with his siesta. I couldn't decide whether to accept this rather forlorn "bird in the hand" or take a chance on another hotel that looked, if possible, even less inviting. Out of the blue, the clerk blurted, "Make up your mind. After all, time is money!" How would you reply to that at a time of day when literally nothing was happening? I couldn't help but laugh at the incongruity of it all. If there ever was a case of time not being money, it was in Patras during siesta in the summer.

Though M-time cultures tend to make a fetish out of management, there are points at which M-time doesn't make as much sense as it might. Life in general is at times unpredictable; and who can tell exactly how long a particular client, patient, or set of transactions will take. These are imponderables in the chemistry of human transactions. What can be accomplished one day in ten minutes, may take twenty minutes on the next. Some days people will be rushed and can't finish; on others, there is time to spare, so they "waste" the remaining time.

In Latin America and the Middle East, North Americans can frequently be psychologically stressed. Immersed in a polychronic environment in the markets, stores, and souks of Mediterranean and Arab countries, one is surrounded by other customers all vying for the attention of a single clerk who is trying to wait on everyone at once. There is no recognized order as to who is to be served next, no queue or numbers to indicate who has been waiting the longest. To the North European or American, it appears that confusion and clamor abound. In a different context, the same patterns can be seen operating in the governmental bureaucracies of Mediterranean countries: a typical office layout for important officials frequently includes a large reception area (an ornate version of Lorenzo Hubbell's office), outside the private suite, where small groups of people can wait and be visited by the minister or his aides. These functionaries do most of their business outside in this semipublic setting, moving from group to group conferring with each in turn. The semiprivate transactions take less time, give others the feeling that they are in the presence of the minister as well as other important people with whom they may also want to confer. Once one is used to this pattern, it is clear that there are advantages which frequently outweigh the disadvantages of a series of private meetings in the inner office.

Particularly distressing to Americans is the way in which appointments are handled by polychronic people. Being on time simply doesn't mean the same thing as it does in the United States. Matters in a polychronic culture seem in a constant state of flux. Nothing is solid or firm, particularly plans for the future; even important plans may be changed right up to the minute of execution.

In contrast, people in the Western world find little in life exempt from the iron hand of M-time.[2] Time is so thoroughly woven into the fabric of existence that we are hardly aware of the degree to which it determines and coordinates everything we do, including the molding of relations with others in many subtle ways. In fact, social and business life, even one's sex life, is commonly schedule-dominated. By scheduling, we compartmentalize; this makes it possible to concentrate on one thing at a time, but it also reduces the context.[3] Since scheduling by its very nature selects what will and will not be perceived and attended, and permits only a limited number of events within a given period, what gets scheduled constitutes a system for setting priorities for both people and functions. Important things are taken up first and allotted the most time; unimportant things are left to last or omitted if time runs out.

M-time is also tangible; we speak of it as being saved, spent, wasted, lost, made up, crawling, killed, and running out. These metaphors must be taken seriously. M-time scheduling is used as a classification system that orders life. The rules apply to everything except birth and death. It should be mentioned, that without schedules or something similar to the M-time system, it is doubtful that our industrial civilization could have developed as it has. There are other consequences. Monochronic time seals off one or two people from the group and intensifies relationships with one other person or, at most, two or three people. M-time in this sense is like a room with a closed door ensuring privacy. The only problem is that you must vacate the "room" at the end of the allotted fifteen minutes or an hour, a day, or a week, depending on the schedule, and make way for the next person in line. Failure to make way by intruding on the time of the next person is not only a sign of extreme egocentricism and narcissism, but just plain bad manners.

Monochronic time is arbitrary and imposed, that is, learned. Because it is so thoroughly learned and so thoroughly integrated into our culture, it is treated as though it were the only natural and logical way of organizing life. Yet, it is *not* inherent in man's biological rhythms or his creative drives, nor is it existential in nature.

Schedules can and frequently do cut things short just when they are beginning to go well. For example, research funds run out just as the results are beginning to be achieved. How often has the reader had the experience of realizing that he is pleasurably immersed in some creative activity, totally unaware of time, solely conscious of the job at hand, only to be brought back to "reality" with the rude shock of realizing that other, frequently inconsequential previous commitments are bearing down on him?

Some Americans associate schedules with reality, but M-time can alienate us from ourselves and from others by reducing context. It subtly influences how we think and perceive the world in segmented compartments. This is convenient in linear operations but disastrous in its effect on nonlinear creative tasks. Latino peoples are an example of the opposite. In Latin America, the intelligentsia and the academicians frequently participate in several fields at once—fields which the average North American academician, business, or professional person thinks of as antithetical. Business, philosophy, medicine, and poetry, for example, are common, well-respected combinations.

Polychronic people, such as the Arabs and Turks, who are almost never alone, even in the home, make very different uses of "screening" than Europeans do. They interact with several people at once and are continually involved with each other. Tight scheduling is therefore difficult, if not impossible.

Theoretically, when considering social organization, P-time systems should demand a much greater centralization of control and be characterized by a rather shallow or simple structure. This is because the leader deals continually with many people, most of whom stay informed as to what is happening. The Arab fellah can always see his sheik. There are no intermediaries between man and sheik or between man and God. The flow of information as well as people's need to stay informed complement each other. Polychronic people are so deeply immersed in each other's business that they feel a compulsion to keep in touch. Any stray scrap of a story is gathered in and stored away. Their knowledge of each other is truly extraordinary. Their involvement in people is the very core of their existence. This has bureaucratic implications. For example, delegation of authority and a buildup in bureaucratic levels are not required to handle high volumes of business. The principal shortcoming of P-type bureaucracies is that as functions increase, there is a proliferation of small bureaucracies that really are not set up to handle the problems of outsiders. In fact, outsiders traveling or residing in Latin American or Mediterranean countries find the bureaucracies unusually cumbersome and unresponsive. In polychronic countries, one has to be an insider or have a "friend" who can make things happen. All bureaucracies are oriented inward, but P-type bureaucracies are especially so.

There are also interesting points to be made concerning the act of administration as it is conceived in these two settings. Administration and control of polychronic peoples in the Middle East and Latin America is a matter of job analysis. Administration consists of taking each subordinate's job and identifying the activities that go to make up the job. These are then labeled and frequently indicated on the elaborate charts with checks to make it possible for the administrator to be sure that each function has been performed. In this way, it is felt that absolute control is maintained over the individual. Yet, scheduling how and when each activity is actually performed is left up to the employee. For an employer to schedule a subordinate's work for him would be considered a tyrannical violation of his individuality—an invasion of the self.

In contrast, M-time people schedule the activity and leave the analysis of the activities of the job to the individual. A P-type analysis, even though technical by its very nature, keeps reminding the subordinate that his job is not only a system but also part of a larger system. M-type people, on the other hand, by virtue of compartmentalization, are less likely to see their activities in context as part of the larger whole. This does not mean that they are unaware of the "organization"—far from it—only that the job itself or even the goals of the organization are seldom seen as a whole.

Giving the organization a higher priority than the functions it performs is common in our culture. This is epitomized in television, where we allow the TV commercial, the "special message," to break the continuity of even the most important communication. There is a message all right, and the message is that art gives way to commerce—polychronic advertising agencies impose their values on a monochronic population. In monochronic North European countries, where patterns are more homogeneous, commercial interruptions of this sort are not tolerated. There is a strict limit as to the number as well as the times when commercials can be shown. The average American TV program has been allotted one or two hours, for which people have set aside time, and is conceived, written, directed, acted, and played as a unity. Interjecting commercials throughout the body of the program breaks that continuity and flies in the face of one of the core systems of the culture. The polychronic Spanish treat the main feature as a close friend or relative who should not be disturbed and let the commercials mill around in the antechamber outside. My point is not that one system is superior to another, it's just that the two don't mix. The effect is disruptive, and reminiscent of what the English are going through today, now that the old monochronic queuing patterns have broken down as a consequence of a large infusion of polychronic peoples from the colonies.

Both M-time and P-time systems have strengths as well as weaknesses. There is a limit to the speed with which jobs can be analyzed, although once analyzed, proper reporting can enable a P-time administrator to handle a surprising number of subordinates. Nevertheless, organizations run on the polychronic model are limited in size, they depend on having gifted people at the top, and are slow and cumbersome when dealing with anything that is new or different. Without gifted people, a P-type bureaucracy can be a disaster. M-type organizations go in the opposite direction. They can and do grow much larger than the P-type. However, they combine bureaucracies instead of proliferating them, e.g., with consolidated schools, the business conglomerate, and the new superdepartments we are developing in government.

The blindness of the monochronic organization is to the humanity of its members. The weakness of the polychronic type lies in its extreme dependence on the leader to handle contingencies and stay on top of things. M-type bureaucracies, as they grow larger, turn inward; oblivious to their own structure, they grow rigid and are apt to lose sight of their original purpose. Prime examples are the Army Corps of Engineers and the Bureau of Reclamation, which wreak havoc on our environment in their dedicated efforts to stay in business by building dams or aiding the flow of rivers to the sea.

At the beginning of this chapter, I stated that "American time is monochronic." On the surface, this is true, but in a deeper sense, American (AE) time is both polychronic and monochronic. M-time dominates the official worlds of business, government, the professions, entertainment, and sports. However, in the home—particularly the more traditional home in which women are the core around which everything revolves—one finds that P-time takes over. How else can one raise several children at once, run a household, hold a job, be a wife, mother, nurse, tutor, chauffeur, and general fixer-upper? Nevertheless, most of us automatically equate P-time with informal activities and with the multiple tasks and responsibilities and ties of women to networks of people. At the preconscious level, M-time is male time and P-time is female time, and the ramifications of this difference are considerable.

In the conclusion of an important book, *Unfinished Business,* Maggie Scarf vividly illustrates this point. Scarf addresses herself to the question of why it is that depression (the hidden illness of our age) is three to six times more prevalent in women than it is in men. How does time equate with depression in women? It so happens that the time system of the dominant culture adds another source of trauma and alienation to the already overburdened psyches of many American women. According to Scarf, depression comes about in part as a consequence of breaking significant ties that make up most women's worlds. In our culture, men as a group tend to be more task-oriented, while women's lives center on networks of people and their relations with people. Traditionally, a woman's world is a world of human emotions, of love, attachment, envy, anxiety, and hate. This is a little difficult for late-twentieth-century people to accept because it implies basic differences between men and women that are not fashionable at the moment. Nevertheless, for most cultures around the world, the feminine mystique is intimately identified with the development of the human relations side of the personality rather than the technical, cortical left-brain occupational side. In the United States, AE women live in a world of peoples and relationships and their egos become spread out among those who are closest to them by a process we call identification.[4] When the relationships are threatened or broken or something happens to those to whom one is close, there are worries and anxieties, and depression is a natural result.

Polychronic cultures are by their very nature oriented to people. Any human being who is naturally drawn to other human beings and who lives in a world dominated by human relationships will be either pushed or pulled toward the polychronic end of the time spectrum. If you value people, you must hear them out and cannot cut them off simply because of a schedule.

M-time, on the other hand, is oriented to tasks, schedules, and procedures. As anyone who has had experience with our bureaucracies knows, schedules and procedures take on a life all their own without reference to either logic or human needs. And it is this set of written and unwritten rules—and the consequences of these rules—that is at least partially responsible for the reputation of American business being cut off from human beings and unwilling to recognize the importance of employee morale. Morale may well be the deciding factor in whether a given company makes a profit or not. Admittedly, American management is slowly, very slowly, getting the message. The problem is that modern management has accentuated the monochronic side at the expense of the less manageable, and less predictable, polychronic side. Virtually everything in our culture works for and rewards a monochronic view of the world. But the antihuman aspect of M-time is alienating, especially to women. Unfortunately, too many women have "bought" the M-time world, not realizing that unconscious sexism is part of it. The pattern of an entire system of time is too large, too diffuse, and too ubiquitous for most to identify its patterns. Women sense there is something alien about the way in which modern organizations handle time, beginning with how the workday, the week, and the year are set up. Such changes as flextime do not alter the fact that as soon as one enters the door of the office, one becomes immediately locked into a monochronic, monolithic structure that is virtually impossible to change.

There are other sources of tension between people who have internalized these two systems. Keep in mind that polychronic individuals are oriented toward people, human relationships, and the family, which is the core of their existence. Family takes precedence over everything else. Close friends come next. In the absence of schedules, when there is a crisis the family always comes first. If a monochronic woman has a polychronic hairdresser, there will inevitably be problems, even if she has a regular appointment and is scheduled at the same time each week. In circumstances like these, the hairdresser (following his or her own pattern) will inevitably feel compelled to "squeeze people in." As a consequence, the regular customer, who has scheduled her time very carefully (which is why she has a standing appointment in the first place), is kept waiting and feels put down, angry, and frustrated. The hairdresser is also in a bind because if he does not accommodate his relative or friend regardless of the schedule, the result is endless repercussions within his family circle. Not only must he give preferential treatment to relatives, but the degree of accommodation and who is pushed aside or what is pushed aside is itself a communication!

The more important the customer or business that is disrupted, the more reassured the hairdresser's polychronic Aunt Nell will feel. The way to ensure the message that one is accepted or loved is to call up at the last minute and expect everyone to rearrange everything. If they don't, it can be taken as a clear signal that they don't care enough. The M-time individual caught in this P-time pattern has the feeling either that he is being pressured or that he simply doesn't count. There are many instances where culture patterns are on a collision course and there can be no resolution until the point of conflict is identified. One side or the other literally gives up. In the instance cited above, it is the hairdresser who usually loses a good customer. Patterns of this variety are what maintain ethnicity. Neither pattern is right, only different, and it is important to remember that they do not mix.

Not all M-times and P-times are the same. There are tight and loose versions of each. The Japanese, for example, in the official business side of their lives where people do not meet on a highly personalized basis, provide us an excellent example of tight M-time. When an American professor, business person, technical expert, or consultant visits Japan, he may find that

his time is like a carefully packed trunk—so tightly packed, in
fact, that it is impossible to squeeze one more thing into the
container. On a recent trip to Japan, I was contacted by a well-
known colleague who had translated one of my earlier books.
He wanted to see me and asked if he could pick me up at my
hotel at twelve-fifteen so we could have lunch together. I had
situated myself in the lobby a few minutes early, as the Japanese
are almost always prompt. At twelve-seventeen, I could see his
tense figure darting through the crowd of arriving business
people and politicians who had collected near the door. Following greetings, he ushered me outside to the ubiquitous black
limousine with chauffeur, with white doilies covering the arms
and headrests. The door of the car had hardly closed when he
started outlining our schedule for the lunch period by saying
that he had an appointment at three o'clock to do a TV broad-
cast. That set the time limit and established the basic parameters
in which everyone knew where he would be at any given part
of the agenda. He stated these limits—a little over two hours—
taking travel time into account.

My colleague next explained that not only were we to have
lunch, but he wanted to tape an interview for a magazine. That
meant lunch and an interview which would last thirty to forty
minutes. What else? Ah, yes. He hoped I wouldn't mind spending time with Mr. X, who had published one of my earlier books
in Japanese, because Mr. X was very anxious to pin down a
commitment on my part to allow him to publish my next book.
He was particularly eager to see me because he missed out on
publishing the last two books, even though he had written me
in the United States. Yes, I did remember that he had written,
but his letter arrived after the decision on the Japanese publisher had been made by my agent. That, incidentally, was the
very reason why he wanted to see me personally. Three down
and how many more to go? Oh, yes, there would be some
photographers there and he hoped I wouldn't mind if pictures
were taken? The pictures were to be both formal group shots,
which were posed, and informal, candid shots during the interview, as well as pictures taken with Mr. X. As it turned out,
there were at least two sets of photographers as well as a sound
man, and while it wasn't "60 Minutes," there was quite a lot of

confusion (the two sets of photographers each required precious seconds to straighten things out). I had to hand it to everyone— they were not only extraordinarily skilled and well organized, but also polite and considerate. Then, he hoped I wouldn't mind but there was a young man who was studying communication who had scored over 600 on an examination, which I was told put him 200 points above the average. This young man would be joining us for lunch. I didn't see how we were going to eat anything, much less discuss issues of mutual interest. In situations such as these, one soon learns to sit back, relax, and let the individual in charge orchestrate everything. The lunch was excellent, as I knew it would be—hardly leisurely, but still very good.

All the interviews and the conversation with the student went off as scheduled. The difficulties came when I had to explain to the Japanese publisher that I had no control over my own book —that once I had written a book and handed it in to my publisher, the book was marketed by either my publisher or my agent. Simply being first in line did not guarantee anything. I had to try to make it clear that I was tied into an already existing set of relationships with attached obligations and that there were other people who made these decisions.[5] This required some explaining, and I then spent considerable time trying to work out a method for the publisher to get a hearing with my agent. This is sometimes virtually impossible because each publisher and each agent in the United States has its own representative in Japan. Thus an author is in their hands, too.

We did finish on time—pretty much to everyone's satisfaction, I believe. My friend departed on schedule as the cameramen were putting away their equipment and the sound man was rolling up his wires and disconnecting his microphones. The student drove me back to my hotel on schedule, a little after 3 P.M.

The pattern is not too different from schedules for authors in the United States. The difference is that in Japan the tightly scheduled monochronic pattern is applied to foreigners who are not well enough integrated into the Japanese system to be able to do things in a more leisurely manner, and where emphasis is on developing a good working relationship.

All cultures with high technologies seem to incorporate both polychronic as well as monochronic functions. The point is that each does it in its own way. The Japanese are polychronic when looking and working inward, toward themselves. When dealing with the outside world, they have adopted the dominant time system which characterizes that world. That is, they shift to the monochronic mode and, characteristically, since these are technical matters, they outshine us. As will be seen later, the French are monochronic intellectually but polychronic in behavior.

4 High and Low Context Messages

Computers have captured the fancy of the Western world. They are marvelous servants for the mind, they can relieve the mathematician of much of the tedious routine work associated with his craft. They simulate all sorts of complex processes and procedures. They link information networks and memory banks, unifying entire countries; they run subway systems and trains, fly the space shuttle, smooth out traffic flow during rush hours, write letters, collect bills, and in fact do many of the tasks formerly assigned to lower and middle management. There is one thing, however, that computers do not do well: translate! This deficiency is not for want of money, need, interest, talent, or brains. Millions of dollars spent on computer translations of Russian after years of effort demonstrated that the most efficient and the most effective translator of scientific Russian is a human being, a scientist. Scientific linguists found that it was particularly important that the scientist have a deep knowledge of the field being translated. The failure of the computers was not in the proper analysis of syntax (grammar) and vocabulary, a monumental task in itself, but in the relationship of the linguis-

tic code to the larger setting of the scientific field: the context
in which each word, sentence, and paragraph was set.

Words and sentences have different meanings depending on
the context in which they are embedded. The word "man," for
example, means one thing when spoken in the context of stages
of maturity for males, another when speaking of units of work
such as "man hours" or "man days" of labor, and still another
when referring to a 100,000-man army (which now includes
women).

Regardless of whether it is ultimately possible for humans to
evolve computer translation to the point where it will be useful, the matter of context will be an issue in any communication
between human beings. "Contexting" seems to be a function
of the right and left hemispheres of the brain,[1] each working
in quite a different way, but in tandem; each supplying an essential element of virtually any communication. No communication is totally independent of context, and all meaning has an
important contextual component.[2] This may seem obvious, but
defining the context is always important and frequently difficult.
For example, language is by its very nature a highly contexted
system. As stated earlier, language is not reality; it is, however,
rooted in abstractions from reality. Yet, few people realize how
dependent the meaning of even the simplest statement is on
the context in which it is made. For example: A man and a
woman on good terms, who have lived together for fifteen or
more years, do not always have to spell things out. When he
comes through the door after a day in the office, she may not
have to utter a word. He knows from the way she moves what
kind of day she had; he knows from her tone of voice how she
feels about the company they are entertaining that night.

In contrast, when one moves from personal relationships to
courts of law or to computers or mathematics, nothing can be
taken for granted, because these activities are low context and
must be spelled out. One space inserted between letters or
words on a computer where it does not belong can stop everything in its tracks. Information, context, and meaning are bound
together in a balanced, functional relationship. The more information that is shared, as with the couple referred to above,

the higher the context. One can think of it as continuum ranging from high to low.

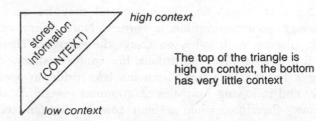

Pair this triangle with another in a balanced relationship. In
this second triangle there is very little information at top and
more at the bottom.

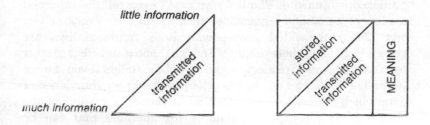

Combine the two and it can be seen, as context is lost, information
must be added if meaning is to remain constant. The complete
relationship can be expressed in a single diagram; there can be
no meaning without both information and context.

The matter of contexting requires a decision concerning how
much information the other person can be expected to possess
on a given subject. It appears that all cultures arrange their
members and relationships along the context scale, and one of
the great communication strategies, whether addressing a single
person or an entire group, is to ascertain the correct level of contexting of one's communication. To give people information they
do not need is to "talk down" to them; not to give them enough
information is to mystify them. The remarkable thing about
human beings is that they make these adjustments automatically

and, in the majority of cases, they manage to produce a high proportion of messages that are appropriate. Here, too, the rules vary from culture to culture, so that to infer by the level of contexting that "they" do not understand may be an insult, even though your assumption is correct. North Germans, for example, place a high value on doing things right. Their approach is of the detailed, meticulous, low context variety. When learning a foreign language, Germans take pride in speaking correctly and following the rules of grammar exactly. It comes as a blow, therefore, when a high context Parisian corrects them.

Examples of how the rules of context work can readily be seen in advertising, particularly in automobile ads in the United States. Contrast the information in advertisements for the German-made Mercedes-Benz with those for the Rolls-Royce produced in England. The Rolls ad won't even tell the informed readers of automobile magazines such as *Road & Track* or *Car and Driver* the rated horsepower; Rolls representatives are known to have replied simply, "Enough," not a sufficient answer for such informed readers, some of whom might even buy a Rolls. The Mercedes ad, on the other hand, has an abundance of data which is devoured by potential buyers. The readers of the above magazines expect a number for anything that can be measured, and horsepower is one of the first numbers they look for. Don't ask me why, but many Americans don't seem to be able to evaluate the performance of anything unless they can attach a number to it. Needless to say, differing readerships require differing amounts of contexting. In the American journal *Science* (the organ of the American Association for the Advancement of Science), there are two types of articles: moderately contexted feature articles for the generalist, and short, high context technical articles for specialists, in which it is taken for granted that the reader knows the field well enough so that little or no explanation need accompany descriptions of experiments.

The communication strategies of the United States and Japan provide a different perspective in the matter of contexting. Americans lacking extensive experience with the Japanese (particularly older Japanese who have not adapted to European

communication patterns) frequently complain of indirection—
they have difficulty knowing what the Japanese are "getting at."
This is because the Japanese are part of a high context tradition
and do not get to the point quickly. They talk around the point.
The Japanese think intelligent human beings should be able to
discover the point of a discourse from the context, which they
are careful to provide. There are several interesting and important spinoffs from this basic difference in communication
strategy. The United States, having its roots in European culture which dates back to Plato, Socrates, and Aristotle, has
built into its culture assumptions that the only natural and
effective way to present ideas is by means of a Greek invention
called "logic." The Japanese see our syllogistic method and its
deductive reasoning as an effort to get inside their heads and
do their thinking for them. It was for this as well as other reasons that many European missionaries had poor results in Japan
compared to other cultures.

On one of my visits to Japan many years ago, only one of the
Jesuit missionaries I interviewed was having any success and
this was because he had learned the difference in the contexting
patterns. He concluded quite correctly that the way to reach
the Japanese was not with the low context logical reasoning of
Thomas Aquinas, but by emphasizing something else—namely
the wonderful feelings that one had if one were a Catholic.
Feelings are rooted in one part of the central nervous system,
while words and logic are a function of an entirely different
area of the brain. To reach people, you have to know which part
of the brain to involve![3]

Another important and closely related feature of both context
and communication is the fast and slow message continuum.[4]
All cultures seem to settle on particular spots on the "message
velocity spectrum" where their members feel most comfortable.
It does very little good to send a message in the fast category
to people who are geared for a slow format. The content of the
two messages may be the same, but once a person is "tuned
in" on a given frequency, messages on any other frequency
might as well not exist. It's like changing channels on a television set, only in this case few people know that there is any
channel except the one they have used all their lives.

What are some examples of fast and slow messages and the characteristics of each? The following chart lists a few:

FAST– Low	SLOW– High
Prose	Poetry
Headlines	Books
A communiqué	An ambassador
Propaganda messages	A well-researched position paper
Cartoons	Etchings
TV commercials	TV series-in-depth
Sports events	History and meaning of sports
Lust	Love
Hollywood short duration marriages	Successful marriages of any sort
"First-naming" in the United States	"First-naming" in Germany
A reproduction of a work of art	Work of art itself
Michael Korda's manipulations[5]	Genuine friendship

This rather cursory list provides an idea of how the pattern operates in AE cultures. Actually, everything human beings consider or pay attention to, as well as most things they do not pay attention to, can be placed somewhere along the fast/slow message spectrum. The meaning of life is a message that releases itself very slowly over a period of years. Sometimes there aren't enough years for the message to unfold. Arnold Toynbee used to study the rise and fall of civilizations for the messages they carried over hundreds and even thousands of years. History falls into the very long, slow category. In the United States, personal relationships and friendships tend to be somewhat transitory. In many other cultures, friendship takes a long time to develop; however, once established, it persists and is not taken lightly. Sex in the United States is treated as fast, when, as a matter of fact, it takes years to develop a close, deep relationship in which sex frequently plays an important part.

The velocity spectrum relates directly to our topic of communication and particularly communication across cultural boundaries. The problem with American television is that the commercial message must be released in a minimum of 10 to 15 seconds or at most 60 seconds. Six commercials during a two-minute break is not unusual. The use of a fast message, by implication, suggests that the product itself is not going to last very long. If this is so, a 60-second commercial would be appropriate only for such items as soap, cigarettes, Kleenex, disposable diapers, perishable vegetables, and household cleaners. It would be unsuitable for banks, insurance companies, Mercedes-Benz or Rolls-Royce, and many small Japanese cars, all of which we expect to be endowed with an aura of permanence. TV commercials are best adapted to fast, low context messages. Diplomacy, statesmanship, and an examination of life, love, and the pursuit of happiness are best done in the high context slow modes, such as books or BBC television series. Buddha, Confucius, Muhammad, Christ, Shakespeare, Goethe, and Rembrandt all produced messages that the world is still deciphering hundreds of years later. The message any spoken or written language gives us is very slow. The same can be said of culture. Human beings have been studying languages for over four thousand years, and we are just beginning to discover what language is all about. Culture, particularly primary culture, will take longer.

In the marketplace, selling provides us not only with interesting examples of fast and slow methods, but also with high and low context. Consistent with the American practice of establishing relationships with others rather quickly, American businessmen tend to think in short-time intervals. In American business, planning intervals range from the quarter to the year. Five-year planning is virtually unheard of in many industries, even though overall strategy may require it. Sales representatives are expected to get results in a hurry. Contrast this with France, where a French colleague who used to be one of my graduate students discovered that, to sell his product, he had to establish long-term personal relationships with his customers, typical of high context polychronic behavior, and that at times it was a matter of selling as many as three generations in a family-owned business. This process could take up to two years and is a highly contexted routine. When his French company was bought by an American firm and an American manager was installed, his new American boss failed to understand why my friend couldn't just walk through the door and sell his product to the client in one or two visits. The American boss was simply not willing to allow the time for his salesmen to develop the proper relationship to sell his products. It is easy to see why customers in France do not belong to the company but rather to the salesman, and why they follow the salesman whenever he changes jobs. A similar long-term pattern exists in Latin America, where people depend more on human relationships, which they consider permanent, than they do on the wording of a contract, which is not. Needless to say, the North American system puts American business at a disadvantage, and only a few managers know why.

There are businesses in the United States that have been built with personal relationships, such as the personalized bookstore where the proprietor knows not only the books but also the customers. The Francis Scott Key Book Shop in Georgetown, Washington, D.C., is an example. Once when my wife was ill, I asked the proprietor to help me select some books for her. Not only did she know what my wife had read but what she hadn't read and was thinking of reading next! Needless to say, such service, while once common in the book trade, is increasingly rare. Mass merchandizing of books is now threatening the livelihood of many of the nation's small bookstores and some book representatives are being replaced with computers. The ultimate effects of moves of this sort are not known. By removing the human element, the personal contact, the feedback from book dealers, and the accumulated expertise of the book sales representatives, decisions are made on the basis of a computer printout or the accountant's sharp pencil. This "bottom line" approach influences which ideas are disseminated and which are not—just one more example of decisions which are moving us further down the road to depersonalization. All societies depend for their stability on feedback from the people. Depersonalization reduces feedback to a minimum, contributing to instability and lowering the overall level of congruence in the society.

Congruence of communication in form, function, and message is a recognizable and necessary element in everything. It is especially easy to observe in art. In fact, it is the basis of all great art, and not nearly enough is known about it. Lack of congruence makes people anxious or ill at ease. The principle can be seen in almost any activity. An example is the individual who is passing himself off as someone he is not. Carefully studied details are there, but if they are not put together in the right way or are used incorrectly, they are incongruent. Music is particularly sensitive to congruence. According to Leonard Bernstein, Beethoven's work has the highest congruence of that of any composer.[6] Beethoven would work on a single note until he got just the right one. When this choice was made, it was clear that no other note would do. Bernstein expressed it somewhat as follows: "It looks as though it's phoned in from God! Every note is perfect." That is congruence; it holds you in awe.

It is axiomatic in AE cultures that the "law" is a permanent feature, one of our most important institutions. Contracts are sacred and binding because they are backed by law. Yet a contract can be drawn up overnight. In Latin America, as stated earlier, only human relationships are regarded as "permanent" and, as such, are treated as slow messages. A North American, therefore, who quite naturally puts his faith in contracts and treats human relationships as rather fast, ephemeral, and not to be depended on, not only will be vulnerable, but will introduce dissonance and incongruence into the situation. Anxiety—when it is a product of incongruence—is not easy to detect and, if detected, its cause can be hard to pin down. Thus, the North American businessman, feeling secure because he has made the proper offerings to his own gods, will assume quite naturally that he has done everything he should to ensure his economic survival in Latin America. Instead, he and his company are a patchwork of vulnerabilities directly traceable to failure to ensure the future by developing the proper networks of friends in the right places.

The types of vulnerabilities I am referring to have nothing to do with size, wealth, geography, or political or military power, but they can be just as crucial in determining the outcome of a transaction. For the reasons just stated, it is not to be expected that the vulnerability will be seen in its true light. In chapter 7, I describe briefly one such vulnerability: that of the American business person confronted with retroactive French fiscal regulations that make people guilty of offenses that were perfectly legal at the time they occurred.

Still another kind of vulnerability directly traceable to differences in time systems is suffered by American business people when competing with the Japanese. Unfortunately, there is little likelihood that American businesses will change their ways, not because they don't want to do well, but because most businessmen are narrow and unsophisticated. From a narrow "hard-nosed" point of view, primary level culture is most commonly seen as trivial. There is no way to provide enough contexting for the overseas American to see the consequences of his decisions. In the twenty-five years I have been working with business, the practice has been to consult the experts only after mistakes have been made—when it's too late.

Another development which some consider trivial is the effect on the family of cultural differences in house type. American-style houses in Japan are now being blamed for weakening the highly cohesive structure of the Japanese family,[7] which in turn contributes to violence on the part of the young. The reader may wonder why I bring up the subject of houses when we are talking about time. The answer is that time and space are inevitably functionally interrelated. In an earlier study, *The Hidden Dimension*, my partner and I found that in rehabilitating houses occupied by the urban (predominantly black) poor, the provision of a room in which children of school age could shut the door and study produced noticeable improvement in their grades. What the space provided was time to be alone with the books without distractions, and it took spatial adjustments to achieve that end. In the Japanese case, what was taken away in the shift from the traditional Japanese home to the American mass-produced bungalows was the time that adults and the growing young spent together. Takeo Matsuda, a successful housing "tycoon" who helped to bring affordable American-style houses to Japan, is now afraid that these houses may have contributed to the rising violence in Japanese families and schools. She says: "Japanese families are very tight. We all studied together and slept together." The American home compartmentalizes the family, so children grow up leading "separate lives" and, as a consequence, lack training in having consideration for others. Matsuda states that now "we have everything," but "we don't help each other."

There can be no doubt that the compartmentalization of time and space have reinforced narcissistic trends engendered elsewhere and are at least in part responsible for the "me first" pattern in American culture. This does not mean that everyone should immediately do away with all partitions. It does mean that Americans would be well advised not to disregard the effect of temporal and proxemic patterns on our lives.[8]

The case cited below is from New Mexico and illustrates the following points: 1) the difficulties encountered by poor people in adjusting to a complex society everywhere; 2) the importance of understanding the role of time as used by such marginal families; 3) the kind of attention and shepherding many poor people need which can only be provided by committed, resourceful individuals from their own group; 4) the almost unbelievable range of differences in a complex culture such as that of the United States; and 5) how sometimes some government programs designed—with the best of intentions—to help people have the opposite effect.[9]

In the United States, ethnic diversity has been with us since the beginning. One of the most recent groups to assume importance in many ways, including sheer numbers, is the Spanish. In New Mexico, Spanish-Anglo contact dates back more than one hundred and fifty years. The two groups have lived side by side, worked together, intermarried, and governed the state in tandem, yet have retained separate identities in spite of their intermingling.

Of all of the problems faced by working-class Spanish trying to adjust to the dominant Anglo culture, the most difficult and basic are those of polychronic individuals having to adjust to a monochronic culture.[10] These points are illustrated by the story of a Chicana friend who established a private school to help families of "battered" children. Describing how the condition of her clients affected the school schedule, she said: "We have to go out and get them, because they can't plan far enough ahead to catch the bus." Drawing a deep breath, she continued, "So that we won't have to spend too much time waiting for them to get ready, I have to call up some of the mothers one hour before our bus arrives—every day of the week. I ask, 'Have you gotten the kids up? Have you gotten breakfast? Have you dressed yourself? Because, remember, you have to come, too.' "[11] Think for a minute about the differences between families in the same country: in one family one cannot even plan ahead for the whole day's meals, but must run to the store for the food for each meal when people become hungry. Contrast that with a family where one member, usually the mother, not only markets for the week or the month, but also creates a complex series of interlocking schedules of events covering months at a time. Unfortunately, most government aid programs are implicitly built around middle-class white American time models. By taking people as they are, it would not be too difficult to create a variety of graded models, incorporating increasing complexity and durations against which progress could be measured. One could even reward families for being able to make the transition successfully from planning one meal to planning and marketing for all the meals for the day!

Moving from intracultural differences, intercultural differences are just as great, but considerably more complex. Consider the implications for Americans working in Ecuador, where the pattern is polychronic. Ecuadorians[12] don't measure their time the way we do. For us, marking time is a little bit like writing a sentence on the page and keeping the words evenly spaced so that they are not crowded. North Americans try to distribute time evenly and if it seems there will be a crunch at the end, for whatever reason, we increase the tempo and the effort to make the deadline so that it all comes out even. The Ecuadorians might know that something has to be done by the end of the day, but they will not act as though anyone should speed up or make any additional effort at all. Photographs that must be processed are left unprocessed. If they stop and talk to someone on the street corner and discover that a mutual friend is in the hospital, everything stops and takes on a new direction, for, "We must go to the hospital and see him!" The North American notion that obligations in time must be honored in order to keep from causing stress to others is a complete and utter non sequitur. The North American has internalized the schedule;[13] Ecuadorian schedules are externalized, and therefore carry very little weight. The internalized network of friends and relatives and attending to their needs takes precedence. Internal controls are much more binding than the external ones because you are your own tyrant.

For more than twenty-five years I have been intimately aware of the difference between Colonial Iberian-Indian (CII)[14] time and AE time. Not only is one polychronic and oriented toward people and the other monochronic, stressing procedures (procedures are fast while people are slow), but there are other differences as well which run from South America to northern New Mexico.

The differences in the way in which time is handled are known to both groups, but neither takes the other seriously. I can still conjure up images of high-ranking, high-speed North Americans in a slow, people-oriented setting, getting hotter and hotter under the collar waiting as the time that they were to be received by some local functionary arrived, passed, and then faded into history. High blood pressure, disenchantment, and ulcers were the most common by-products of North Americans working in the CII area. Why? The answer did not come immediately, but had to be teased from the matrix of behavior at home in the United States over a period of years. Once identified, the reasons were obvious.

To understand the pattern that was making Americans so unhappy overseas, it is first necessary to say a few words about the difference between external and internalized controls. In Westerners, the conscience is the most familiar and widespread example of internal control mechanisms. I do not believe that it is the only one, but it may be one of the most important. Failure to follow the dictates of the conscience leads to guilt and/or anxiety. I feel anxious when I see I am going to be late and will do what I can to avoid it, not only because I feel an obligation to be on time but because I want to avoid anxiety. Since my time controls are internal, I do not need to be told or reminded that I should be on time. The CII pattern has a different set of loyalties and controls. First, their obligation, as suggested earlier, is not to the schedule, but to the people they are with, particularly to those who are related or close. So what happens when people are inevitably late or fail to show up? The individual waiting is not upset because there are so many things going on that it usually doesn't matter that much. Besides, being kept waiting is not read as a slap in the face. The schedule is something that is outside and is not linked to either the ego or the superego as it is for those reared in North European cultures. Being late is not a message nor is it taken personally. For the North American, however, time set aside for an activity that doesn't happen is wasted and can never be reclaimed or recovered. But that's not all.

AE time isn't just structured functionally and used to control the flow of work, activities, and involvement with others, but how it is handled is also deeply *symbolic*. Time is not only money, but also a symbol of status and responsibility. The amount of leeway an individual has in a given time system is a signal to all indicating where that individual is placed in his organization. It isn't just "the name on the door and the Bigelow on the floor" that counts, but how much discretion a person has in the way he can schedule time. Even within the confines of a single culture, there are problems because the patterns are seldom spelled out. They demonstrate, however, how subtle and binding PLC patterns can be. For example, some individuals will unknowingly violate these unwritten mores (thereby making things difficult for themselves and others) by using the primary level systems to communicate a status that they have not earned —the equivalent of putting on airs. For a subordinate to treat time in the same way that his boss does, taking liberties with the system (days off, arriving at irregular hours, taking off early, and generally stretching the system to the limit) can lead to tremendous resentment as well as anxiety among fellow workers. People work many years for these privileges and flexibilities in their schedules, and it makes them resentful to see someone who is a subordinate taking such liberties with the system.

This is somewhat comparable to cutting in front of people in line. Waiting one's turn is one of the basic structure points of AE culture. Those who have waited the longest or been around the longest are seen as being the most deserving. The time system and what people make of it gets even more involved with the status connotations of different waiting times in offices. While there are regional and ethnic differences, the pattern for the majority—particularly on the Eastern Seaboard of the United States—is that the passage of waiting time in an office carries three sets of interrelated fast messages to the individual who is kept waiting: first, there is the importance of the business itself, second is status, and third is politeness. If Blevins, who sits behind the closed door, feels his caller, Wood, has important business, he will make sure it is handled expeditiously at the appointed time; likewise, if Wood arrives half an hour early and expects to be seen, he may be putting Blevins down by making his own business seem more important than it really is. On the other hand, if their business is really important, then Blevins will have cleared his desk in advance when his secretary announces: "Mr. Wood from Mutual of Nebraska is here to see you, Mr. Blevins." (Wood is here to discuss a long-term loan that Blevins has been negotiating for months.) Under conditions of this sort, Blevins will be ready to go whenever Wood arrives. To fail to do so would communicate disinterest or casualness that might make a lender look elsewhere.

Nevertheless, for routine matters, if a waiting time of twenty-five to thirty minutes occurs, this long period signals that the status of Wood's business has been symbolically reduced to nothing and that boundary has been crossed into the region of rank and status. It is bad enough to have his business discounted, but then that just may be the situation. Blevins, because of matters beyond his control, is simply not interested or can't act right now. However, unless he wants to be cruel or rude, he will not keep Wood waiting to a point where serious inroads on his status have been made. It is at this point that Wood's blood pressure usually will begin to rise as he becomes more and more upset.

If the reader is a member of the AE culture and is not either consciously or unconsciously rebelling against the system, he should have little trouble recalling instances when he was kept waiting long enough for it to begin to reflect on his status. How one handles such a situation is purely personal. Some physicians and most hospitals are, of course, notorious in their violations of the nation's mores. It would not be so bad if it weren't so stressful for the patients. The message comes through, however, that hospital staff is in charge; they are important, the patient isn't (he's lucky to be there). The patient's status, by all applicable nonverbal criteria and measures, has in most instances become reduced to that of a nonperson. It wasn't designed that way; it just happened. But it doesn't have to stay that way. Our hospitals are even harder on polychronic people than they are on the monochronic ones. At least for AE peoples the schedules are understandable, even if what is being communicated by those schedules and the people who make them is hard to swallow. What polychronic people need, which they can't have in our hospitals, is lots of friends and relatives around them all the time. But regulations and scheduled visiting hours won't permit this.

Whenever controls are external, as occurs with industrial time clocks and schedules in the military, schools, business organizations, and hospitals, one finds that they are frequently rigidly enforced. It all depends, however, on the administrator. The problem lies in the inequities and exceptions, because not everyone can get to work at the same time. The writing of regulations to handle all contingencies is virtually impossible and can become cumbersome. Such complexity builds up and adds to the burdensomeness and size of government. The general rule is that externalized controls inevitably lead to complexity. It is almost impossible to simplify from the outside, because usually only the people who are directly involved are in a position to simplify anything.

A similar pair—closely related to time—which confounds matters in almost any office where both monochronic AE and polychronic CII people are interacting, is how these two groups are oriented toward people and procedures. This is not simple, and there are ramifications that go beyond what one might imagine. Whenever it is necessary to get action, to change something, to adjust the system to prevent past mistakes from reoccurring, the automatic response of the AE group is to take a look at procedures, and then to try to develop suitable alternatives to correct matters. Yet, it is this procedural orientation that makes our bureaucracies so vulnerable to fraud. All that is needed is the correct "procedure," and the bureaucratic wheels start grinding.

The CII group's response is quite different. First, if there is a breakdown of any kind they immediately see who is involved and what the human consequences will be of taking remedial action. If it is a personal matter, a more responsive system is needed and the question will be: "Who do I know who can help me?" If it is an organizational matter, they find who is in charge and who can help them. If they don't know someone at that moment, they will look until that connecting link in the human chain is located. Obviously, systems employ both people and procedures, but the emphasis in each cultural group is different.

It is almost impossible for fast-message, procedure-oriented people to bypass their procedures—even in an emergency. They have the feeling not only that it is not right, but also that without the proper procedures nothing works. Remember that external controls tend to become complex—which means that the procedures of the CII people-oriented cultures can be cumbersome. This invokes a double penalty on the poor AE bureaucrat stationed in Rio who gets a cable from New York to be on the next plane to Washington. He has an emergency need to clear the country on three hours' notice, when Brazilian "procedures" require three or more days. Brazilians, of course, simply find a friend and bypass the procedures.

It is sometimes difficult for North Americans in Brazil to realize that one of the costs of doing business is taking the necessary time to cultivate friendly working relationships with individuals strategically located in business and government. Until this is done, nothing will happen. Who one knows and has as a friend can make a lifesaving difference in an emergency. A Latin American does not (cannot) simply act on the abstract assumption that the individual who has been waiting in line longest is more important than his friend. If one has to wait,

it simply indicates that either he is not connected (and therefore
not worth bothering with), or he doesn't know how to get along
with others and has no friends. Either of these is an indictment.

Procedures, schedules, and linear, monochronic time cluster
just as orientation toward people in bureaucratic settings, polychronic time, and rather loose schedules cluster. Having worked
at the interface of the AE and CII cultures for most of my life,
I have discovered again and again that PL culture is "for real."
It isn't just that certain CII individuals cannot be depended
upon to arrive at the appointed time or cannot complete projects
according to AE schedules, but that these people are in the grip
of one pattern, while I am in the grip of an entirely different
one.

5 Culture's Clocks:
Nuer, Tiv, and Quiché Time

Winston Churchill once said, "We shape our buildings and then they shape us." This was during the debate on the form that Parliament should have when it was rebuilt following the bombing of London during World War II. Churchill was right, of course, but it isn't just space that shapes us; time does the same thing, only it's hard to realize that when we build our time systems we are shaping our lives. It is only recently, however, that time and space have begun to be recognized as influencing the direction as well as the outcome of behavior.

E. E. Evans-Pritchard, authority on the African peoples and one of the grand old men of anthropology, was the first to my knowledge to have a major section in an ethnography devoted to "time and space."[1] He also recognized that there was something inherently different about the way in which the African people actually experienced and structured time. Realizing that time could be different from one culture to the next, he had sufficient insight to accept it as a product of the human mind and not as a constant transcending culture. Evans-Pritchard even categorized the time system of the Nuer (a sub-Saharan tribe) in two ways: eocological (sic) and structural. Eocological time was essentially time dealing with seasons, annual cycles, movements of animals, in fact, all the temporal aspects of nature in the Nuer cosmology. In addition, the Nuer structured time as a component of cultural and social life. Discussing this, Evans-Pritchard observed that "the subject bristles with difficulties." He was right, of course. The subject does "bristle with difficulties," many of them as yet to be confronted. It is a tribute to Evans-Pritchard's sensitivity and perceptiveness that he recognized this. Speaking of his experiences, he states: "With the Nuer . . . *time is not a continuum but a constant between two points* . . . the first and last persons in a line of descent." Nuer time was fixed as a channel through which kin and groups moved. The visual metaphor evoked by his analysis is like a stroboscopic picture of a turning wheel when the spokes are frozen in time by the synchronized flashes of the light. In Nuer time one knew the wheel was moving, but it had the appearance of standing still as generations fed in at the hub slowly worked their way up to the top of the spoke at the rim. The Nuer realized that time moved, in a sense, but for their purposes it was necessary to treat it as though it did not—for them only the generations moved.

My colleague Paul Bohannan,[2] following in Evans-Pritchard's footsteps, made the next insightful record of African time while working with the Tiv of Nigeria. Bohannan went much further than Evans-Pritchard in describing how Tiv time works, for which we are greatly indebted to him. To the Tiv, time was somewhat like a series of enclosed rooms, each containing a different activity. The walls of time, like the hollow conduits of the Nuer, seem to have been relatively fixed. "Time rooms" could not be moved about or shuffled, nor was the activity in those rooms to be changed or interrupted, as occurs in AE cultures.[3] Once inside one of their time-activity chambers, both the Tiv and their activity were inviolate. Like the time clock in a vault, they were sealed in and safe from interruptions. This is a situation Westerners may envy, because even Presidents are not immune from interruption.

Dominated as it is by activities, Tiv time is intimately associated with the famous African markets. Markets, like everything else that humans produce, have a regional character. The

Tiv market is restricted to a single category of commodity; the
Tiv know what day it is by what is being sold in the market, i.e.,
leather, brass, cattle, vegetables, etc. Each area, as might be
imagined, has its own sequence of produce and commodities so
that the weekdays occur in a different sequence. There is no
attempt on the part of the Tiv to integrate these differences, for
it is quite evident that for the Tiv time provides a permanent
frame in which activities fit like the pieces of a jigsaw puzzle.
Viewed from the perspective of a Westerner, it must be reassur-
ing to have *time, space,* and *activity* not only fixed but also
congruent. With a time pattern such as this, with markets and
goods forming a permanent fixture (like the wheel of Nuer
time), relationships develop and form between merchant and
buyer as they do the world over. Each day imprints the people
with its special character, produced by the tempo and personalities of the market. Each new week provides a reinforcing
replay of the preceding weeks. It is not difficult to recall somewhat analogous frames in the United States, for example, the
woman who takes every Thursday to go to town, or the weekly
or monthly rounds of a traveling salesman, but neither of these
is related temporally to a centralized institution like the market
in which everyone participates, nor are the personalities in the
orbit necessarily fixed as they are with the Tiv.

To some Westerners it must sound as though the Tiv are in a
rut, knowing what to expect from one week to the next. In our
own culture this could happen, but there is a difference. AE
peoples create ruts, but it is done in a very interesting way. We
call them routines. Routines are seen in the repertoires of transactions, the responses evoked from others and ourselves. We
know we are in a rut when we foresee what someone else is
going to say next and what our reply will be. We're in a rut
when there are no surprises, nothing is new, particularly no
new ideas, and when there is no new development in literature,
art, music, dance, or when there are no new ways of looking at
things. H. H. Munro (Saki) wrote a delightful satire on this.
His protagonist, Clovis, visits an English clergyman and his
housekeeper-sister, who are definitely "in a rut." Overhearing
their complaints of ennui, Clovis arrives at a correct diagnosis
and then proceeds to make a shambles of the clergyman's life

and his household by totally disrupting all the routines. The title Munro chose for this little story was "The Unrest Cure."

Barbara Tedlock's *Time and the Highland Maya*[4] is an unusually rich description of how the Quiché Indian (pronounced "*key chay*") time is organized and how it leads to an experience of living which is totally different from that of AE cultures. Translating from Quiché time to AE time is possible only in the most limited sense.

The Quiché Indians, descendants of the Maya and occupants of highland villages in Guatemala, inherited the Maya calendrical system, one of the most advanced in the world at the time of the Spanish conquest of Mexico and Latin America. The Maya recorded the lunar and solar cycles, calculated eclipses, as well as the orbits of Venus, Mars, and Jupiter, with an accuracy equal to and sometimes superior to that of their conquerors.

Not only were religious festivals accurately timed, but the daily lives of the people were so intertwined with the time system that it is impossible to understand one without the other. There are even specialists—diviners—in interpreting Quiché time. The term for diviner is Ajk'ij, meaning "day keeper." To get at the core of Quiché culture, anthropologist Tedlock and her husband apprenticed themselves to a diviner and are now practicing Quiché diviners. Because day keeping is a sacred function, diviners are actually shaman-priests. In fact, the entire Quiché curing-religious system is built around divination and the "day keepers," who are also the people's link to the gods.

Following the tradition of their ancestors, the Quiché have two calendars, one civil and the other religious (sacred-divinatory). Composed of 18 twenty-day months, the civil calendar year totals 360 days with 5 days remaining. There are 260 days to the religious calendar, which has no months but is an assemblage of 20 combinations. These two calendars interlock like two rotating gears to produce the Calendar Round, which only repeats itself once every fifty-two years.

Not only are there two separate calendars that mesh, but there are other qualitative pattern differences worthy of attention. In the United States and Europe it is generally taken for granted that there will be a beginning and an end to everything: saying good-bye to the old year and ushering in the New Year; the capital at the beginning of a sentence and the period at the end; the first day of the week and the last day of the week. Name anything you wish: relationships, jobs and careers, meals, and working for advanced degrees—for everything there is a beginning and an end. Yet, in the Guatemala highlands, Tedlock provides convincing evidence that the Quiché 260-day calendar, like the wheel which it resembles, has no beginning and no end.

There are other differences. To the Maya, the sacred 260-day calendar provides the base on which an elaborate system of divination is built. Each day has special characteristics and it takes a shaman-diviner to provide a proper interpretation of the day. This is particularly important when critical decisions are contemplated. Not only does each of the twenty days have a proper name and character that is divine, but also a number. The "nature" of the days change depending on the numerical accompaniment, as well as the actions or moves contemplated during that particular day. A "good" day in one context may be bad in another. There are favorable and unfavorable combinations, and it is the combination that determines how a day should be interpreted.

The very nature of the calendrical system and people's relationship to it forces them to ponder each day to decipher its sacred message, a message relating to each life in a different way. Mayan calendars are structured in such a way as to direct people's attention not only to interpreting the significance of the various combinations of names and numbers but also to what these combinations mean in different contexts. The context in which the message is set in this way becomes part of the meaning. Need I add that the Quiché are a highly contexted people?

Contrast the above with AE calendrical systems. Our days have no formal differences except for Sunday, the Sabbath, and holidays like Thanksgiving, Easter, Christmas, and New Year's Day. All of our days, even the holidays, are virtually indistinguishable. The days of the week are the same regardless of the month, season, or year. The etymology of four of the days of the week indicate, of course, that these were originally named after the Scandinavian mythological gods: Tiw's day, Woden's day, Thor's day, Fria's day, which may have given them an entirely different significance in days past.

To the average European, however, the Quiché system sounds like astrology. This is *not* the case, but, again, one must take into account the context. AE people look at astrology as the opposite of science. Ergo, for many Europeans astrology is not taken seriously because it is not scientific. My point about the Quiché is not whether an astrological type of system is valid or not, or logical according to Western definitions. Rather, I'm concerned about the effect such a system has on the people— how they think and live. The Quiché calendar and what goes with it represent a mental environment in which the people spend their entire lives.

Also, Quiché time performs specific functions related to curing, religion, economics, well-being, marriage, family, and village welfare. The ancestors and mythological beings, sorcery and witchcraft are all involved and reinforced by the time system.

Given the task of describing a time system such as the Quiché Mayan in words poses problems. The words assume more importance than they should. Words, after all, are symbols, and while it is the symbols that are used to describe what the people do, somehow in this process the symbols and the story they tell take on a life of their own. This creates a new reality that is quite different from the reality expressed by the Guatemalan Indians. As readers—and writers—we are lacking in adequate contexting. I am trying to remind the reader that this reality about which I speak exists as something distinct from what I or anyone else says or thinks. The Quiché reality causes them to scrutinize each day and its character as it relates to their own character, their desires, and their past, as well as the tasks that lie ahead. The Quiché really do have to think deeply and seriously about the process of how each day is to be lived. Our time system has quite the opposite effect.

With AE peoples time is an empty container waiting to be filled; furthermore, the container moves along as if on a conveyor belt. If time is wasted, the container on the belt slips by only partially filled and the fact that it is not full is noted. We are evaluated by how those containers look. If they are all full, that is a strong plus. If they are full with good deeds and creative productions, then we can feel we have lived a "full and productive life"! Judged by this standard, some people are seen as more productive than others and require bigger containers while the rest of us sit back in awe of how much they accomplish in their lifetime. To have done little or nothing means no containers are filled. Sitting around passing the time of day with others, incidentally, is in the "nothing" category. Yet, there are people who judge by other standards, lead very productive lives simply by being encouraging, helpful, and supportive of others. These good souls—and they are good souls—are sometimes made to feel that they haven't "made much of life" because other people's containers are full but where are theirs?

Compared to cultures like the Quiché, ours seems unusually self centered because our time system keeps reminding us that we are the only ones who can fill those containers. Our own unwritten rules tell us other people cannot help. Time itself is seen as neutral and its only value is that it is relentless and unfeeling; it waits for no man.

For us the law of productivity still applies. Americans must make every moment count, because each container is divided into hours, minutes, and even into seconds. We look back and say, "I can't imagine where the week went," or "It's Monday and the next thing you know it's Friday and the week is over" —a roundabout way of saying, "I didn't get as much accomplished as I thought I should have." The Quiché don't feel they have to make every moment count. The Quiché face a different, more subtle task: how to use each day properly.

For the Quiché, living a life is somewhat analogous to composing music, painting, or writing a poem. Each day properly approached can be either a work of art or a disaster if the proper combinations are not found. For individuals brought up in the AE tradition, these differences are not easy to articulate or understand. Why? Because we pay so little attention to what it means to live right. In our part of the world, living is something that is taken for granted. It's done automatically. Living has a lot to do with filling those containers—with meeting objectives.

On the group level, Quiché time has been a great source of strength in helping the people in the difficult and sometimes almost impossible task of integrating alien European institutions, material culture and customs in ways that make sense to them. It is the Quiché way to *consciously* evaluate everything from the outside world. If the item is judged to be beneficial, it is then adapted to local custom; otherwise it is rejected. As a result, nothing is ever felt to be alien or strange. Unwanted, alien customs, beliefs, procedures, and ceremonies are simply ignored. In that way the foundations of life are not threatened. In this regard, the difference between our two cultures has far-reaching consequences. In the United States, blacks and Native Americans are like islands surrounded by a sea of AE culture. Over the years both have suffered varying degrees of destruction of their institutions, some of which might have been avoided had they been lucky enough to have shared some of these crucial Quiché patterns of adaptation or rejection.

It is necessary to digress a moment to say a few words about what happens when the members of two cultures meet. Some cultures tend to be more receptive to what is different, while others are less so. The AE group has trouble coming to terms with anything different. As a consequence, there are strong, deep currents of proselytization very close to the mainstream of all AE cultures. We are the ones who send missionaries not just in religion but in virtually every aspect of life. Americans more than most seem dominated by the need to shape other people in our own image. This drive to clone our own culture is accompanied by the implicit conviction that culture is something that one dons and doffs like a suit of clothes. Unlike many other peoples on the globe, we don't experience our own culture as something that penetrates every cell of our bodies, which is the source of all meaning in our lives. Since culture is seen by Americans as something superficial which can be shed without disturbing what lies underneath, we are frequently blind to the disastrous consequences of our addiction to proselytizing.

Students of culture now generally agree that black informal (out-of-awareness) primary culture survived the ravages of contact with whites during slavery. For Native Americans the history is both brighter and sadder. On the bright side, the

Pueblo Indians of the Southwest have managed to survive with
most of their culture intact. Other Indians were less fortunate;
their cultures either were crushed or collapsed, because they
proved to be extraordinarily vulnerable to AE customs and
institutions. In this respect the Quiché are quite remarkable.
They are well defended against Europeans and their culture.
There may be a clue to how this immunity to cultural disintegra-
tion works in the way in which time is handled.

According to Tedlock, the Quiché treat time as a dialectic,
which means, in her terms, that "at no given time, past, present,
or future, is it possible to isolate that time from the events that
led up to it and which flow from it." How different from
Americans, who discard the old and eagerly clasp the new to
our breast. We see this in our attitudes toward ideas, books,
music, automobiles, styles, and, most recently, marriage partners.
Even when we rediscover the old, it is treated as new, like the
move to search for one's "roots." We live in a culture where
most things are disposable; continuity simply is not there. Also,
whenever anything new is incorporated or adopted—a belief,
a life-style, or even a spouse—there are deep, unconscious patterns that make us feel we must automatically disavow the old.
In disavowing our past, we fragment history and in the process
manage to break the few remaining threads that bind, stabilize,
and give unity to life.

The Quiché do not have this problem, and while neither culture planned it that way, the results are there for us to examine.
Given the dialectic nature of the Quiché time system (which
knits everything together), the Quiché do not feel the need to
rid themselves of the foreign elements already integrated. This
principle was applied when Tedlock was unhesitatingly accepted
as an apprentice diviner on the same basis as a Maya—something that would never have happened among the Pueblo
Indians. Quiché thought, as molded by Quiché time, did not
require that she disavow her own culture, only that she integrate
the new material into what was already there. Under similar
circumstances in the United States, we would resolutely try to
"cast out" the alien elements regardless of how long they had
been around or how deeply integrated they were into the psychic
structure of the individual.

We can see this disavowing pattern in the whole "born-again" syndrome of American life. The reality is that it may be possible to alter one's view of one's past or events in that past, but the past is still there; it will never go away and it cannot be altered.[5]

To summarize the significance of Quiché time, the day-keeping divination is a constant reminder of: A) the sacred time system; B) one's obligation to one's lineage as well as bonds to that lineage reaching back in time (the indiscretion of an ancestor can be the reason behind today's misfortune); C) one's relatedness to the earth and its nature spirits and to the gods; and D) one's relationships with and obligations to the larger community. We have very few ceremonies in AE cultures where one is forced to think about god, family, and one's relationship to self and others.

Quiché time binds people to the village, to ancestors, gods, and daily life. It all comes together in the divination process, at which point another connection is added, tying people into their physiology. An important feature of Quiché divination is the use of the body as a sender, receiver, and analyzer of messages. I do not refer to "body language" but to physiological functions of the body itself, which is read the same way one reads a book. The Quiché view the circulation of the blood as an active agent in the body's message system. Ability to divine with the "blood" is said to be a direct gift of the ancestors. Either the shaman's own body or the patient's body can be used. If the patient's body is used, the shaman feels the pulse in different parts of the body. This "pulsing" is not like our own, where the rate is used diagnostically; rather, the total character of the pulse is analyzed for its meaning, including twitching and tingling. In one instance, the shaman sat with her legs straight in front of her. If the pulse moved up the inside of the legs, the patient would live. If it moved up the outside, the patient would die. In addition to the blood, the shaman makes use of feedback from the long muscles of the body which twitch, contract, or tingle. Where the blood is experienced as a sensation and how the muscles respond tell the story.

Again, the Western reader may treat this type of diagnosis as hogwash—simple superstition. In our frame of reference this

would be right, because AE peoples know very little about how
to read the messages of the body. Our knowledge is limited to
slow messages of the psychosomatic type and leaves out the fast
ones that signal what is happening right now. Tedlock and her
husband learned to read their own bodies the Quiché way and
they state categorically that it provided a variety of feedback in
their relations to others that had been heretofore unknown to
them.

   In another context, I once knew a psychoanalytic colleague
who saved his life by depending on this very same system
to tell him what was going on in a patient. This therapist was
treating an unusually attractive, seductive, and assaultive female
patient. Having narrowly escaped death from attacks by the
patient on two occasions, my friend decided that a more reliable
way of staying in tune with his patient was required. A split-
second ducking reflex once prevented his head from being
crushed by a weighted smoking stand. Would he be that lucky
the next time? The attacks came without warning and with
lightning speed. There were none of the usual external signs
that something was about to happen. All of the sensory cues on
which people depend for feedback were absent. In fact, it
seemed almost as though the assaults occurred when least expected and when the therapist was relaxed and most vulnerable.
Those therapists who have had experiences with assaultive
patients also know that it is important to be able to control
one's own level of anxiety, and this is not easy if one may be
struck at any moment by a heavy object. Looking for a solution,
my friend discovered that unbeknown to his conscious mind his
body had been picking up signals before an attack. In his own
pulse rate there was a highly reliable, built-in, early warning
system, which he then proceeded to monitor regularly. A quick-
ening beat sent the unmistakable message: "Look out!"

   There are references in the literature on the Maya to the
twitching of muscles, but until I read Tedlock it never occurred
to me that these could be anything but unexplained aberrations
in the body's rhythm system. Later in this book I will devote a
chapter to body synchrony and to what one body tells another
when the two move in synchrony. In *Beyond Culture* I referred
to these mechanisms as the body synchronizers. Although there

are some very suggestive leads, it is still not known exactly how these messages are sent and received. It would appear that the Quiché shamans may have developed or elaborated on this process and their own knowledge of how the system works, or enhanced their awareness of their own bodies as senders and receivers of messages (probably all three). Whatever the explanation, the result is heightened awareness of an important component of human consciousness. In light of the current interest in consciousness raising I would say that it never pays to dismiss something simply because there is no adequate explanation of an experience that is only partially understood.

On the basis of almost fifty years' experience with cultures covering a very wide range of complexity, I am convinced that the West has made a great mistake in writing off the very special knowledge and abilities of the rest of the world simply because they don't conform to our standards for scientific paradigms. There is still much to be learned from the proper study of other cultures.

6 The East and the West

Insights gained from examining primary level systems have proved so rewarding that I think it is worthwhile to discuss two of the world's leading countries from this point of view: Japan and the United States. My aim is not so much to increase understanding or to diminish misunderstanding (which may be even more important), as it is to enhance appreciation of the underlying cultural processes—to motivate people to be more inquisitive about those things they take for granted. I would also like to communicate some of the tremendous possibilities that lie ahead if the human race can be weaned from its fascination with technology and turn its attention once more to the study of the human spirit. The material on which this chapter is based is drawn from my own experiences as an anthropologist,[1] from Zen Buddhism, and from some of the better known books on Japanese culture.[2]

Since World War II, when Ruth Benedict wrote her landmark book, *The Chrysanthemum and the Sword,* contact between the United States and Japan has increased on a massive scale. Japan's success in American and European markets and a shrinking globe have had the combined effect of greatly increasing the demand for relevant material on the two cultures. The quality as well as the quantity of published material has improved immeasurably in the intervening years. However, there is one element lacking in the cross-cultural field, and that is the existence of adequate models to enable us to gain more insight into the processes going on inside people while they are thinking or communicating. We need to know more about how people think in different cultures, as well as how they organize and explain ideas. What is perceived and what is left out? What is an "idea" or a "concept" as defined by a Japanese compared to an American? What is important? How are ideas organized? According to what principles? How are the separate events that go to make up an idea organized? Some of the answers to such questions can be gleaned from ascertaining where a given culture is on the context scale. Is it high or low context? It is most important to learn how time is structured.

It was not too surprising for me to discover that cultural time is one of the keys to understanding Japan. To begin with, Japanese time, Zen Buddhism, and the concept of MA are all intimately interrelated—relationships which are sometimes difficult for a Westerner to understand. In making this observation, I am not saying that understanding the Western mind is any easier for the Japanese. Remember, time as I have been using it is a core system in our lives around which we build our picture of the world. If the time systems of two cultures are different, everything else will be different. As I stated at the beginning of this book, I do not accept the Western notion that time is an absolute. One studies how cultures handle, experience, use, and talk about time as a way of gaining additional insights into those cultures and understanding the psychology of the people.

In many respects, Zen has represented the ultimate enigma for Americans. The koan "What is the sound of one hand clapping?" is just one example. Koans are sayings or instructions to disciples which appear on the surface to be illogical or impossible, but which have a deeper meaning underneath. To understand a koan it is necessary to understand the context. Westerners find koans difficult to understand because we think of Zen as a concept, a philosophy, or a religion. It is none of these. According to those authorities who write about Zen[3] it is a "way," and a rather extraordinary way at that. Zen represents one of the basic means by which people learn. In Zen we find an excellent example of what I have termed informal learning—learning which depends almost entirely on the use of models, practice, and demonstration.[4] Words are anathema to Zen because words distort. It is not hard to see the stumbling blocks presented by Zen for a culture like ours that begins everything with a question and is constantly asking, "Why?"

Zen is extraordinarily high on the context scale, probably as high as it is possible to get. This means that Zen communication is incredibly fast. One of several reasons Americans have trouble understanding Zen is simply that we are not properly contexted about the history of Zen. We don't know the background of all the koans. This is illustrated by an example taken from a little book by Yoel Hoffmann, *Every End Exposed: The 100 Koans of Master Kidō:* "Master Anzan went into the mill to see Master Sekishitsu. He said, 'It is not easy is it?' Sekishitsu replied, 'What is so hard about it? You fetch it in a bottomless bowl and take it away in a formless tray.'" What Sekishitsu was communicating is ". . . that if one is aware of the *mu* ['bottomless,' 'formless'] aspect of things, one can take them for what they are . . . with this attitude, there is nothing hard in milling or any other work." It is the bottomless and formless shape of things that Westerners find so difficult to accept.

If we only knew in the Western world how much of our lives actually contain within them the seeds of Zen. Unfortunately, many of us spend our lives denying this fact and, as a result, we deny an important part of ourselves. This process of denial interferes with our being able to take the next step—discovery of the hidden energy and power that enables us to do things like draw the great bow with the muscles of the arms and shoulders completely relaxed.[5]

There is some Zen in the way in which the Pueblo Indians of New Mexico teach and interact with each other. An example was provided several years ago by the late John Evans, son of Mabel Dodge Lujan. Evans was superintendent of the combined Pueblo agency and was stationed in Albuquerque about 130 miles south of Taos Pueblo. After a long search, he had finally managed to find an agricultural extension agent to work

at the pueblo who seemed to suit the people and who got along
well. Everything went fine, through the summer and the winter.
But one day in the spring John was visited by this extension
agent, who looked rather forlorn. Standing in John's office, shifting his weight from one foot to the other, he blurted out, "John,
I don't know what's wrong, but the Indians don't like me
anymore."

The following week John drove to Taos and sought out one
of his friends, an old Indian, one of the chief religious leaders of
the Pueblos, and asked him what had gone wrong. The Indian,
playing the part of a Zen master, remained silent. John Evans,
standing there on a spring day warmed by the bright New
Mexico sun, looked back at him. Finally the Indian said, "John,
he just doesn't know certain things." And that was as much as
he would say. Here is the Native American version of Zen
modified slightly by contact with whites, but still incomprehen-
sible to most of them. The Indian wasn't being recalcitrant or
difficult. He knew that John Evans knew the answer and that he
would have to use his head to get it.

Evans' stepfather, Tony Lujan, was a Pueblo Indian from
Taos, where Evans had spent years as a boy and where the
Indians were used to treating him as one of their own. This
explains why he got the answer he did. If it had been someone
else, there would have been no answer at all or else some non
sequitur. After Evans had thought about it for a while, the
answer came with a flash. Of course, how could he be so dumb?
In the spring, Mother Earth is pregnant and must be treated
gently. The Indians remove the steel shoes from their horses;
they don't use their wagons or even wear white man's shoes
because they don't want to break the surface of the earth. The
agricultural extension agent, not knowing all this and probably
not thinking it important if he did, was trying his best to get
the Indians to start "early spring plowing"!

Of course, there is much more to Zen than simply refusing to
give easy answers to the uninitiated. What struck me was that
the patterns were virtually the same in these two very different
cultures. And while the date has not been precisely established,
the forefathers of the Taos Indians must have crossed the Bering
Strait into the western hemisphere somewhere between ten and

twenty thousand years ago. But put that Taos Cacique (religious leader) in the place of the Zen master and John Evans as his apprentice and the dialogue would be virtually interchangeable.

None of this means that a Zen master or a student of Zen is going to understand the Pueblos of New Mexico any more than any other outsider. All non-Pueblo people are almost totally lacking in not only specific information but also context, and the Pueblos want to keep it that way. It only means that if you set up your communications in certain ways, the patterns, if not the content, will be similar. Both the Pueblo Indians and the Japanese are brought up and live much of their lives in close-knit, highly contexted situations. This is why there are no questions, no explanations, and there is great difficulty understanding and accepting outsiders. Yet once accepted, the outsider becomes an insider. If people take the trouble to learn how to work the inside system, it will work as well for them as for anyone else.

Everything that has been said so far applies to time. There is no indigenous philosophical approach such as one finds in the West with its preoccupation with defining what time IS. The American Indians I know have no word of their own for time, and in Japan one doesn't find the extensive preoccupation with this subject. What is Japanese time? How does it compare with Western time? First, it is not and never has been considered an absolute. Time is not imposed on Japanese music as we impose time on our music with a metronome or a conductor. Japanese musicians and their music are "open score." Their music, like their time, comes from within themselves. For example, nema-washi is the term for the time required to get everybody's cooperation as well as consensus. This has been likened by the Japanese to a needle following the groove of a Victrola record. The record is a spatial mechanical metaphor for the process and the transactions that must be accomplished in all parts and at all levels of an organization. The center of the disk where the needle comes to rest symbolizes that the nemawashi process has reached the very top. The speed of the disk is not part of the metaphor. The nemawashi is finished when it is finished and not before. Just like the beginning of the Pueblo Indian dances—they start "when things are ready" and not a moment sooner.

We concentrate on the contrasts between the two cultures,

particularly those which have proved to be stumbling blocks to understanding or to the smooth running of everyday transactions in business and government, and begin with the experiences of Eugen Herrigel, a German professor of philosophy who took up Zen after World War II, with some surprising results. Herrigel's wonderful book *Zen in the Art of Archery* is not only a masterpiece but also a treasure chest of insights on how differently the cultures of Europe and Japan approach virtually everything. Herrigel came to Japan to teach but also to learn. He spent six years studying Zen under a master archer. It would be fruitless for me to attempt to explain to Westerners how one learns a philosophy by shooting a bow and arrow. Herrigel's small volume achieves that purpose more effectively than I ever could. But let's try to put his experience in a somewhat different frame.

Western philosophies are not commonly viewed as paralleling either religion or daily life. D. T. Suzuki makes it quite apparent that Zen Buddhism—which is both philosophy and religion—is actually very close to the underlying primary level culture of Japan. In Japan, the discovery of self is directly linked, therefore, to the full realization of the basic social laws by which one's relatives, friends, neighbors, and countrymen live. In contrast, Western religion, philosophy, and daily life are, following our one-thing-at-a-time scheduling mode, sealed off from each other in tight compartments, while philosophy is a way of training the conscious mind in the search for "truth" or the "meaning of life." In Japan, philosophy is deliberately set up to bypass the word world of the mind and to help the individual tap the wellspring of his own life. There philosophy is life, and it is also the deep core culture of the people.

In the West, archery is sport. In Japan, archery can be a sport, but it can also be a religious-philosophical ritual, a discipline to train the mind. Archery in Western cultures implies the instrumental objective of "hitting the target," which depends on training and strengthening the body. While we are training the body, the Japanese in the Zen tradition follow spiritual exercises designed to expand the mind. When Western people train the mind, the focus is generally on the left hemisphere of the cortex, which is the portion of the brain that is concerned with words and numbers.[6] We enhance the logical, bounded, linear functions of the mind. In the East, exercises of this sort are for the purpose of getting in tune with the unconscious—to get rid of boundaries, not to create them. We follow an "actor, action, goal" paradigm, a manifestation of the grammatical structure of the language: the archer (actor) hit (action) the target (goal). In the practice of Zen archery, the purpose is to blend the archer, arrow, bow and string, and target into a single, unified process. We train by mastering skills; the Japanese train by emptying the mind and eliminating consciousness of the self. In the West, there are schedules that tell us what to do and when; time is an outside force helping us to organize our lives. In the East, time springs from the self and is not imposed. The purpose of Zen is to attune one's self to nature and to "eat when hungry and sleep when tired."

In the West we "organize" our thinking, make plans, theories, and designs for action; we calculate. In Zen, "thought" interferes with consciousness. That is, Zen thinking is natural and unconscious, whereas Western thought is conscious and analytical, leading to dogmas, creeds, and philosophies (content). Zen is more oriented toward context and form. Yet how many archers could duplicate the feat of Herrigel's master, who shot the entire distance of the target hall at night with only a lighted taper illuminating the target area, and split one arrow with another!

In addition to bypassing conscious thought, one of the goals of Zen is to dissolve the ego, to rid the individual of feelings about success, failure, and consciousness of self. To be a master, the Zen swordsman must eliminate all feelings about the dividing line between life and death. Herrigel says: "Every master who practices an art molded by Zen is like a flash of lightning from the cloud of all-encompassing truth. This Truth is present in the free movement of his spirit, and he meets it again, . . . as his own original and nameless essence." Truth in the West is specific, whereas to the Zen master it is all-encompassing and, paradoxically, the very essence of the self.

In the West, we impose our view of nature on man and nature alike because we think of man as separate from nature. Since the early Greek scholars, we have made word pictures of reality in our heads, projected them on the world, and treated these pic-

tures as real. These projections are like the image of gaslight
on the screen of a nineteenth-century Soho stage. Using words
and mathematics, our thinking in the West has been predomi-
nately linear, out of necessity and design. Our thinking is
therefore left brain and low context[7] and ultra-specific. Yet, we
learn from Japan (and, similarly, from Native Americans like
the Hopi[8] and the Navajo) that there is another kind of logic
which complements our dialectic—the logic of *hara*, which is a
logic of context and of action not limited to word paradigms. It
should begin to be obvious that in some of the most basic ele-
mental aspects of life, Japanese and Americans are radically
different. Nowhere is this more evident than in art. Art in Japan
encompasses all of the Zen disciplines, including flower arrangement, archery, and swordsmanship. As a consequence, much of
art is highly contexted.

Four important elements of art—*hara*, MA, intuition, and
michi (the way)—tell us even more. *Hara* links the individual
to nature, since it expresses that part of the person that is
innately and irrevocably natural and an expression of nature
(internalized nature). The Japanese artist, be he a practitioner
of one of the martial arts,[9] potter, painter, actor in a Noh play,
archer, calligrapher, or poet, begins with nature on the inside.
Nature is not something that is outside and separate which he
is trying to reproduce. MA, the second element of art, is a
space-time concept and a meaningful pause, interval, or space.
Silences in Japan shout the deepest feelings. With us, the silence
stands for embarrassment, "dead air," a time in which nothing
is going on. Intuition, the third element, comes from long, deep
study and experience. It is the distilled essence of a theme, an
emotion, idea, or object. *Michi*, the way, implies devotion to
discipline and perfection in one's art. Our closest approximation
to *michi* is technique.

Most Western artists are influenced by two critics. One is
internal, the other external. Whether the artist wills it or not,
there is always a set of aesthetic and visual conventions present
which make up the context in which his work is set. Apart from
the abstract schools is his own need to understand and represent the object before his eyes as faithfully as possible. In contrast, the Zen artist, after years of disciplined exercise, experi-

ences the object with his whole self and then "lets the object draw the picture using the inkbrush as a tool." There is seemingly no conscious effort on the part of the artist to direct the brush. As was true of the Zen archer, the object—the brush— and the artist are part of a single, unified, integrated process.

The difference is that the Japanese, in order to develop his art, must center his efforts on self-knowledge and ultimately on enlightenment. The greatest efforts are made to still the mind and to eliminate the ego, which is subject to the frailties of praise, success, failure, and lack of recognition. Enlightenment is its own reward. The Western artist, on the other hand, with few exceptions, can hardly be likened to a shrinking violet. The ego plays an important role in the life of the Western artist. Those with weak egos have trouble surviving. The vulnerability (if such is the case) of Japanese art is that it grows from within and is less subject to enrichment from the outside. If his own analysis is correct, the Japanese artist would not normally be in the position of learning about his own unstated assumptions when confronted by either radically changing times or a foreign culture. His tendency is either to learn the outsider's system in its entirety or else to turn inward, a characteristic response in other matters as well. The Western artist, while he cannot be counted on to learn a great deal about himself through introspection and meditation while working, does seem to integrate the work of other artists into his own in quite a different way from the Japanese artist. Our artists are much more prone to concentrate either on the aesthetic context, or on the object, or both, than to use art as a way of gaining insight into the workings of their own psyches. They use their art either as a way of expressing what they see, hear, and feel, or as an aid to understanding what they see, hear, and feel. This doesn't mean that the artist never uses art as an avenue to understanding of self, only that it is not in our tradition as an implicit function of art. If it were, we would not see the outraged indignation that occurs in audiences when artistic productions violate the artistic mores of the group. It is as though we in the AE cultures share the outside, while the Japanese and possibly other Eastern cultures share the inside. There are additional differences. Art in Japan is traditionally not compartmentalized life, separate and apart, as it is in the West. It is rather the very essence of life.

It is interesting to note that nowhere in Japanese thought do we find any mention of individual talent. The implication is that virtually anyone who applies himself can become a master in one of the arts and while there are acknowledged "greats" who are given the status of "National Living Treasure," the assumption seems to be that the talent is in the cultural unconscious, rather than in the individual. The Zen approach, of course, places great demands on the individual, while failure is seen simply as insufficient application of discipline, work, and dedication. Failure in the West is frequently chalked up to low aptitude. I am sure that the Japanese recognize aptitude informally, but they don't seem to use its absence as an excuse for poor performance.

An additional difference is that in the West, while large sums of money are paid for the work of big-name artists, those who are not known are apt to have a pretty thin time of it. Nor are Westerners likely to do what I have just done—take art as serious data on the life and mind of our culture. We are rather more inclined to look at economics and politics for insights into the patterns of cultural psychology. Not so in Japan. Swordsmanship, flower arrangement, archery, calligraphy, and art are all viewed as equally valid avenues to understanding the heart and soul of the people and their traditions. In the West, we look for truth in one place. The Zen master knows that enlightenment can be found everywhere.

By now the reader may be saying, "This is all well and good, but who is going to devote his life to the mastery of archery— even in Japan? How typical is this? What proportion of the people are Zen practitioners? None of this seems to explain why the Japanese have been so successful in the world marketplace and why they have managed to wrest leadership in electronics, and the producing of motorcycles and automobiles, from the Western manufacturers."

There is, of course, another side to Japan which is much more visible, which we in the West must also begin to understand. What I have been describing is the underlying bedrock that slowly breaks down into the soil of everyday life. Japanese bedrock and the soil of Japanese culture are comprised of the same constituents, but in different forms. It is the cultural soil of the Japanese and American gardens which I now wish to discuss. As one might surmise, they are quite different.

M. Matsumoto,[10] the Japanese author, interpreter, and translator, states that the Japanese act from three centers: mind, heart, and *hara* ("gut" or "belly"). Because of the highly situational character of Japanese culture, it is important to know which of these three may dominate a given situation. Mind is for business, heart is for home and friends, while *hara* is what one strives for in all things. It should be noted, in regard to *hara*, that while Japanese tradition seems to place great value on it, today, I'm told, *hara* is more commonly associated in people's minds with politicians rather than with Zen masters. The dichotomy between mind and heart has a somewhat different character than in the West. Again, it is in part a matter of situation, but the pull of Japanese culture as a whole is to the heart and not the mind, whereas in the West it is the opposite. To provide an example, because Japan is a high context culture[11] (few rules are stated and a great deal must be filled in with the imagination), maintaining proper personal relationships in business is most important. Contracts won by foreigners with the accountant's sharp pencil are frequently lost later because of neglect of the heart. The heart you can depend on; the mind is always changing. It takes *hara* to integrate the two.

An extension of the above is that in the West we need to understand and appreciate three more things in order to function in Japan: *tatemae* (sensitivity toward others, public self), *honne* (sensitivity toward one's own private self), and *suji* (the situational significance of an event). *Suji* assumes an immersion in a highly contexted culture and includes not only an understanding of the manifest content of a communication, but also an appreciation of the situational contexting aspects. This includes everything else in the situation which has a bearing on the roles that are being acted out.

Sensitivity toward others, while not highly developed in the West, is at least a cultural value acknowledged by some. However, until quite recently, sensitivity toward self has been slighted or looked upon as selfish narcissism, which is far from the case. My impression is that Westerners place much greater stress on the public self and somewhat less on the private self. The contemporary generation has made more progress in changing this than have their parents and grandparents. What is still lacking, however, is the integration that is expressed in *tatemae, honne,* and *hara,* as well as the difference in emphasis on the private and public selves mentioned above.

What about the Americans doing business in Japan? They are very much like business people in other parts of the world in the sense that they tend to take their own way of doing business and their own cultural assumptions for granted and assume that things will work out the way they do at home after one gets some experience. The American abroad—even when he is most successful—is likely to voice such sentiments as "After all, when you get to know them, they are just like the folks back home." Yet one American, an unusually gifted and sensitive man who had worked in an Asian country for almost two decades, learned exactly how to get things done the Japanese way. Commenting on a recent personal coup which received international recognition, he said that he had done his best to understand the culture and that it was important to learn to do things in a "roundabout way which is different from my own country and yet not be phony about it or lose my own identity and become a different person in the process." Part of this "roundabout way" lays great stress on ceremony.

Daily life in Japan is replete with ceremony—young women even bow to greet customers entering the Ginza department stores. But like everything else in Japan there are contrasts which are seen in the great internal drive on the part of the Japanese to move away from formality, to leave behind the public selves (*tatemae*) and move toward the more comfortable, less rigid, warm relationships between private selves (*honne*).

Consistent with this, the Japanese depend more on being able to develop good human relationships than they do on legalistic formality. We in the West demand a carefully worded contract, which we see as our only hold on someone else. Yet, a European approaching business in this manner in Japan is finished before he starts. What do people do to keep from failing? There are institutions to help with this. The evenings spent at nightspots with colleagues and clients are for the specific purpose of finding each other as human beings and establishing bonds of friendship. Not only is friendship accorded more importance than in our country, but when a Japanese makes a friend, he doesn't just drop him later when he is no longer useful, as happens much too often in the United States.

The Japanese take an extremely dim view of anyone who changes his mind or the rules of the game once an agreement has been reached. To fall back on some legal technicality, a policy change, a shift in the political climate, or the thought that a better deal can be made elsewhere, will only make enemies who will take revenge later. You may not even know when it happens.

Virtually all relationships in Japanese culture can be put in one of two classes: close and not close (*honne* and *tatemae*)— us and them! There is nothing in between. Since it is difficult to work with someone over any period of time without establishing a close relationship, one finds that time binds one's self by silken threads to the lives of others. The loyalty of the Japanese to those with whom they are closely tied can be described as nothing less than fierce. On the other hand, while we assume that people will be "on the team" when working together, we have nothing approaching the Japanese dedication and loyalty to the group whatever that group happens to be. Such loyalty is why they are able to depend upon the proper relationship to ensure that things go as planned and why they have no need for contracts. The above also implies a number of other points. When an American in Japan does make up his mind or decides on a certain course of action, he should stick to it. This will not be easy for some Americans, who are in the habit of changing everything at the last minute.

While the Japanese expect to make a profit, their considerations in computing the "bottom line" are much more inclusive than ours. Our "bottom line" is restricted to dollars and cents, while theirs includes an evaluation of possible contributions to national welfare, relationships within the company, networks of people, and much more. If there is any lesson the West could learn from Japan, it would be to expand the concept of the "bottom line" by making it more inclusive—social costs and long-range effects on the country or the market should be considered. And while such a move in a country like ours would not be easy, the overall benefit would be considerable, to say nothing of reducing our tunnel vision syndrome, which makes us especially vulnerable during times of rapid change. Special interests and single-issue politics are two of the most formidable obstacles to be overcome in such a move. The Japanese learned to depend upon each other in ways that we do not, and these dependencies, which are taken for granted, not only strengthen the organization but are part and parcel of the pattern just described.

Related to dependence is the Japanese concept of *giri*,[12] obligations which you incur during a lifetime and which must be repaid. To be in the debt of a rich, powerful person is considered good because that individual can watch out for your interests. Mutual dependence is even better, because each is accumulating and paying off *giri* at the same rate. This would be extraordinarily difficult for Americans, many of whom are averse to being dependent on anyone. We take pride in our independence; in Japan, it is the other way around. It should be noted, however, that dependence in Japan has little or no relation to neurotic dependency as it is known in Europe and America. There is nothing neurotic about the way in which a young, up-and-coming Japanese will use the help of those who are in power to protect as well as to further his interests.

By way of contrast, the act of getting ahead in the United States and in European countries is dependent on being able to stay in the limelight. We seek publicity, to stand out in a group. To see this basic pattern åt work, all one has to do is look at half-a-dozen high school yearbooks. The drive in this country is for recognition. You can see it in stance, dress, posture, attitudes, voice level, and in our possessions. Our idols are public figures. The ones who get the greatest pay in sports, theater, and business are the ones who are best known. All of this works against us in Japan. Americans who want to do well in Japan have to develop a whole new approach based on being unobtrusive and avoiding attracting attention. This takes some doing on our part. The rewards do not go to showoffs in Japan.

THE EAST AND THE WEST

Japanese who work overseas pay a heavy price for their absence from home. If one is away from Japan or out of touch, it weakens his ties to others. However, there is also a lesson here for the American who does business in Japan. One can build on this dependence if one takes it seriously and does not become too wrapped up in squeezing the last nickel out of every transaction. Many a contract lost on price has later been picked up again because of this need to be constantly in touch. Neglect of the customer on the part of the low bidder can turn victory into defeat.

In the United States, we strive for a meeting of minds; in Japan, it is a meeting of hearts. In the United States, top management makes the decisions; in Japan, the real decisions are made at the middle level. Start at the middle and when things are right you will reach the top. In the United States, we spar with words to show who is smartest. In Japan, people synchronize their breathing. With us, differences of opinion aren't serious; in Japan, they can be very serious, hence the need to avoid confrontations. To be highly articulate in the West is an advantage. Not so in Japan, where to be too articulate is a disadvantage and is inconsistent with *hara*. They allow for the MA in rhetoric. It permits people time to think. Timing is everything in both cultures, but the context is much broader in Japan.

7 The French, the Germans, and the Americans

Primary level cultural differences between AE peoples and the Japanese are to be expected. But what about relations between Americans and Europeans? Many Americans learn either German or French in high school and college and certainly tend to think of the two peoples as more like us than Arabs, Hindus, or Malaysians. They would be right because, culturally, white Americans are closer to Europeans than anyone else. After all, most of our ancestors came from one part of Europe or another. However, it so happens that there are unanticipated differences within the AE group, some of which are rather extraordinary.

Few of these differences are as apparent on the surface as those between the AE group and cultures in other groups in the world, but this only makes them more enigmatic, particularly when encountered in daily life. American business is not only particularly vulnerable but also frequently blind to the risks being taken in Europe because significant differences are found in virtually every aspect of life. How can this be?

The most basic of culture patterns are acquired in the home, and begin with the baby's synchronizing his or her movements with the mother's voice.[1] Language and our relations with others build on that basic foundation of rhythm. In the American home, schedules are introduced almost immediately, and in the early twentieth century, schedules were even regulated as to when the baby was nursed. It was schedule first, baby and mother's needs second. This has fortunately changed in recent years. When the child enters school, however, the culture comes on full force. Schools instruct us how to make the system work and communicate that we are forever in the hands of administrators. Bells tell everyone when they must begin learning and when to stop.

More than thirty years after I graduated, I couldn't help being startled, saddened, and at times exasperated by the ear-shattering bells in the halls of the various universities where I taught. Those bells punctuated the beginning and end of each class period, and were completely unnecessary, because both students and professors had years ago internalized the whole process of scheduling. Even the most oblivious and insensitive professor would have a hard time ignoring signals emanating from students when it is time for the class to end. Somewhere at the bottom of the bureaucratic morass there must still be a line item in the budget for the maintenance of bells. The message, of course, is that there is an administration calling the shots. Time is imposed! Internal rhythms, classroom dynamics, effectiveness of learning and teaching are all subordinate to the schedule. Nevertheless, even though administrations dominate our lives, we in the United States, when compared with the French, are relatively decentralized.

In France, until very recently, what was taught and when in every classroom in the land was dictated from a central point —Paris. All periods and all subjects in the French school system were scheduled in advance. At any time of the day it was possible to tell what any child in any city or village was studying. Consistent with this centralized orientation in scheduling, the French have centralized virtually everything else in both time and space. Their bureaucracies are much more powerful than our own, and within French bureaucracies the middle position is considered to be the strategic one. French bureaucracies are also deeply committed to the welfare of France and, I am told, will subordinate their own interests to those of the country.

Regrettably, this is not always the case in the United States. Another difference between France and the United States is that in France businesses—particularly banks—are not in an adversary relationship to the government. Even if there weren't the "old boy" network of classmates from school, it would be unthinkable that business and government would not cooperate with each other.

Centralization permeates France down to and including the individual business person, manager, or executive. It is important, however, to keep in mind that size is another significant variable in any organization. Larger organizations must be more tightly scheduled than smaller ones. A single person at the head of a small office can carry in his head both individual and organizational needs, and this makes time shifts more manageable. The central point—the person in the middle—is the place where everything comes together and from which all power and control originates. The center defines the situation and, as we shall see, the nature of time. This central orientation applies regardless of the size of the organization.[2]

Some of the consequences of this centralized pattern have been cause for much concern on the part of American businesses in France. There are always surprises in store for those who do not know how the French system operates. American logic, business practices, and definitions of fairness seldom apply in France. According to bankers interviewed in Paris and in the United States, most American executives really do their best to do things right and abide by the French requirements. Yet no matter how careful, how meticulous in their reporting of financial transactions, or how conscientious in adhering to agreed-upon plans, it is still possible to wake up one morning to discover that their bankers are faced with retroactive fiscal regulations. The American is dismayed and frustrated because he finds it difficult if not impossible to plan in such an environment.[3]

Unaware that he is in the grip of an immutable time system —the rules of which are not only automatic but also out-of-awareness—the American is outraged when he runs afoul of the French system. Since we have never had to question our own rules and don't know the new ones, we can only say: "They can't do that!" or "It is unfair. What do they mean, retroactive!"

It is clearly impractical as well as inappropriate to try to change the French, so the American has no alternative but to give up treating time as a constant represented by the system he was brought up in, and accept the fact that he is facing a new set of rules, which, like his own, are unstated. Until this crucial step has been taken, it is not possible for him to develop appropriate strategies for coping.

The French in the United States are confronted with a different set of problems. Viewing the social and business world as a set of influence networks, the Frenchman does not at first realize that, unlike France, there is no real center of power in the United States. Certain people have influence, but they are scattered throughout the society and represent different interest groups. Some French business people in the United States, often recent arrivals to this country, give the impression of being vociferous social climbers and snobs, only interested in knowing the right people who can do them some good. All they are really trying to do is locate the true center of power and discover who has influence to ensure that nothing devastating happens without their being prepared. In France, if one does not have a link to the influence networks where financial and other crucial decisions are made, it is possible to become bankrupt overnight. These strategies are necessary because, like everything else in France, core culture time is centralized and the authority to literally turn back the clock lies in the hands of a few individuals in the Ministry of Finance who draft the fiscal regulations on which the welfare of the country depends. The French may tell foreigners in France that they must obey the law and take their financial reporting responsibilities seriously, but they do not warn them of the very important fact that, in France, it is possible to reverse the flow of time. Therefore, it is the responsibility of the foreign businessmen to keep informed as to how the French Government is viewing a constantly changing world. The French attitude is that if outsiders are unable to keep in touch, then they should not be doing business in France. The Frenchman's first loyalty is to France, and from this perspective he couldn't care less what other people think. It is almost as though they unconsciously and continuously restructure the past to justify the present.

The American business person overseas is at a disadvantage for other reasons as well. Americans do *not* put their national interests ahead of everything else,[4] so that business and the government are seldom on the same side working together, and the bickering between competing government agencies as to what our policies are and how they are to be enforced creates a never-never land in which the business person seldom knows where he is. It takes a national catastrophe or an attack on the nation before Americans abandon their own interests and pull together as a team. Under conditions such as these, planning is difficult because no one can agree on a coherent national policy. Time on the organizational level, therefore, is not a ribbon or road to the future but more like a series of small circles with a radius of three months. The midpoint of each circle—the now— is sacred and one does not attempt to move it. The direction in which the circles move depends upon the complex interrelations of all the different competing groups.

But we are not alone. The Germans and the French have trouble at the primary level. Consider the experience of a typical French businessman, M. Chandel (not his real name), working for a German manufacturing firm with overseas subsidiaries. His experience highlights differences in the high and low context systems which occur in various combinations with monochronic and polychronic time, centralized and decentralized lines of authority as well as open and closed score planning.[5] I will try to describe what happens when a monochronic time system is combined with a low context, decentralized social structure. Or when a closed score organization with open score time (French) must incorporate individuals who are used to open score organization and closed score time (German). The differences as well as the repercussions are quite extraordinary.

Results in the real world are deeply influenced by these four interrelated contexts. Unfortunately, there are no metaphors in the English language to express adequately the structural relationships between these quadratic sets and the communication that results from combining the parts in different ways. How then do we talk about these things? One can experience them as operating living wholes, but as soon as we begin to separate and identify the basic components, the unifying patterns dis solve before our eyes. One has the feeling that chemistry might come closer conceptually to what we want to express than the building blocks provided by the spoken language. I mention chemistry because the world of matter, like the world of culture, is composed of a limited number of elements that are combined in different ways. As any metallurgist knows, simply by adding infinitesimal amounts of different substances such as tin, manganese, and cobalt to steel, entirely different characteristics are taken on by the metal. Polychronism is a single way of organizing events in time, and the difference between a polychronic institution and a monochronic institution is like that between night and day. Combine either with highly structured management and the result will be very different. Everything changes.

Chandel is an unusually perceptive manager at the top of the middle level in a German company. He is responsible for the French operations. Chandel is polychronic, as are his French colleagues. He states, however, that the French picture themselves as monochronic, consistent with systematized linear culture in the Cartesian sense. I should explain that there is nothing unusual about people actually being one way but having an entirely different image of themselves. Intellectually and philosophically the French can be monochronic but still be polychronic at the PLC level in the context of daily life—particularly in interpersonal relationships.

Chandel feels most comfortable working in a highly centralized system of control. He feels at ease working in the French mode where "power is in the middle." When asking a superior for a decision in his own polychronic high context, centralized system, he expects a simple "yes" or "no" answer. His expectations are based on the assumption that his superior will know the situation (and thus not need contexting), which is a characteristic of organizations that are centralized and polychronic.

Yet, when M. Chandel addresses the Germans, asking similar questions, he discovers that his German counterpart requires extensive orientation and detailed descriptions as to the nature of the problem, and endless time is required explaining the German position. When this happens, Chandel feels "put down" because they are "low contexting" him by talking down to him.[6] After reading *Beyond Culture*, in which the contexting com-

munication subsystem is described, Chandel began to realize
that what he had taken personally was not meant to be personal
at all, but was simply the difference between a high context
way of communicating and a low context mode. This insight
did not alter the way he felt about being "low contexted," but
it did eliminate his feeling of being "put down," because he
was then able to translate from German to French behavior
and vice versa.

The highly monochronic Germans, with their need for privacy
and order, puzzled Chandel in other ways. The Germans compartmentalized the business and cut themselves off from each
other and this did not make sense to him. The chain of command
on the informal, daily working level seemed not only chaotic
but also quixotic. For example, when working in France, Chandel's boss—the man next in the chain of command above him
in the organization—seemed to have very little real authority
over him (something that would never have happened in his
French culture). Chandel's situation was not unique, except
that few people are as observant and analytical as he happened
to be. The explanation lay in the fact that in his company—and
others like it—there were at least three different authority networks, each with its own information channels: a) the technical
structural organization of the company with its formal chain of
command and associated organization charts; b) a professional
specialized and substantive line of authority with its organic
disciplines as the basis for the information channels—engineers
talking to engineers, chemists taking their cues from chemists,
lawyers attending to lawyers, and all at different organizational
levels; and c) a network of influential people within the company whose success and drive had associated them with highly
productive profit centers. These were men who had reputations
for getting things done. Chandel's links were with "b" and "c"
rather than "a", a situation with which many Americans are
familiar even though it is at the informal rather than the technical, explicit level of culture.

It is no surprise, therefore, that the picture of German organizational structure as seen through French eyes is that virtually
anyone within a given quadrant of responsibility can, if he is

strong, smart, and ambitious, pick up the ball and run with it. This is a pattern that should also be familiar to Americans. German and American cultures are quite similar. The two cultures also have in common a deep commitment to technical organization charts. Both take for granted that the technical organization and the procedures that accompany it may not agree with the informal reality[7] of day-to-day operations.

Another German rule of the informal but binding type is that, once in your organizational box, you not only have the authority to do your job, but no one will bother you. However— and this is important—*you must not make waves!* Much to Chandel's amazement, the German system was not only flexible, but it worked! It permitted great latitude for talent and aptitude, even leeway to the point of tolerating incompetence. As long as the individual didn't cause trouble, complain, or criticize, and was not too obvious about his shortcomings, he was left alone. There was, as might be predicted from the above, great reluctance on the part of the Germans to fire anyone. German eyes turned inward, protected by closed-door, soundproof offices that were structural metaphors of the underlying unconscious cultural facts—a direct expression of the reality of German primary culture.

A note on the American behavioral counterpart: Americans are somewhat hybridized in this respect. In contrast to the Germans, the Americans have an "open-door" policy, and office time is not quite so monochronic. Germans with their drive for order try to fit their entire society into a single temporal plan, whereas the Americans are content to work within the time frame of a single organization. While there are not nearly the number of conflicts between schedules in the United States as one observes in Germany, there are conflicts between the demands of home and office, for example. The implications of all this will become apparent in a moment.

Though ours is an "open-door" policy, American bosses are frequently much less available than they would lead you to believe. As is the case with the Germans, the strong personality who is not in the chain of command but has nevertheless taken over is a common occurrence with us. Anyone who has been in the Army knows about first sergeants and sergeant majors who tyrannize their officers. Bureaucracies of all types are infiltrated with this syndrome—secretaries who won't work, supply clerks who won't get supplies, postal clerks and cloakroom personnel who are rude to the public.

At the time this book was being written, a "classic" case was being reported in the press.[8] According to news reports, two U.S. Food and Drug Administration employees, on their own initiative, had arbitrarily outlawed the inexpensive solutions commonly used for sterilizing soft contact lenses. They had also, without authority, approved a much more expensive saline solution manufactured by a little-known company. The net effect was to increase the sales of that company from $5 million a year to many times that amount. The company was later reported to have been sold to a Swiss company for $110 million. What is extraordinary about this story is that the authority to approve the saline solution did not rest with these two individuals. One of them, according to FDA sources, "wrested decision-making from superiors." Again, parallel to the German pattern, these two bureaucrats who had been wined and dined and received all sorts of favors were not fired when the news broke. Instead, they were put on "leave with pay" (suspended). After almost a year, and only after the second set of hearings by a congressional committee had again publicly established favors and friendship between the benefactors of FDA decisions and the individuals who had made those decisions, the workers were suspended without pay.

This sort of arbitrariness occurs in many countries, but the difference is that in France individuals must be situated in a centralized position before they can tyrannize others. Remember the concierge of years past? Bosses in France are perfectly free to use their subordinates' time in any way they see fit; German bosses don't have that kind of control over employees—the monochronic system and the schedules that go with it are sacred, as we shall see shortly. In Germany it would be unthinkable to routinely ignore an employee's scheduled needs and make him work during the scarce hours set aside by the government when the shops are permitted to open. Additional observations on how the two cultures contrast in organizational settings appear below:

FRENCH   GERMAN

Authority and Control

Bosses exercise authority over subordinates' time. Secretaries are expected to work overtime and on weekends if their bosses need them.

Subordinates' time is sacrosanct, particularly secretaries'. Even on a coffee break they cannot be disturbed, nor can they be kept overtime, for to do so would result in missing limited hours when shops and markets are open.

The combination of polychronic time and the centralized authority makes it impossible for the French to schedule in the way the Germans do.

One thinks of the entirety of Germany as one vast inter-locked schedule. By law, in order to protect employee rights, markets can only be open during certain hours of the day. Office workers must be able to shop at those hours, otherwise the family doesn't eat!

Decision-making

Agendas are more fluid and a function of the situation as it develops.

Answers are expected to be of the decisive "yes" or "no" variety.

In a low context system, infor-mation must be highly struc-tured so that everyone knows where he is. Memos to subordi-nates cover the deep past (back to Charlemagne). Agendas are important and should be ad-hered to. A good manager throws a protective screen around talented subordinates because the open score, informal organization fosters competition that can be destructive. In America, we are less likely to protect subordinates simply because they are good workers; we let them "take care of themselves."

Information and Strategies

The French centralized system favors the decision tree linear pattern, moving from higher centers to and through subordinate centers.

German management is a bit like chess: strong pieces can dominate any level. Like the American counterpart, this system is also subject to great blockages, depending on the individuals involved.

Polychronic relationships require —in fact, demand—strong screening for people in responsible positions. Secretaries and subordinates provide this screening. The French don't like to use the telephone because not only does it deny them information from the face and body, but it makes them too available. Hence, the "Pneumatic," which can get a letter from one part of Paris to another in an hour.

A strong person, while appearing to cooperate, can either run with the ball or block and obstruct, thus killing the initiatives of others below him or frustrating a boss above. The model is reminiscent of the city-states in Europe prior to the Renaissance. There are, however, chances for talent to rise in the system.

French managers have a heavy responsibility to stay informed. Information flows around at a single level of insiders. If you are not part of that group, you may have trouble getting your message through.

Image

In a variety of situations, the French seem to be more prone to reveal who they are than the Germans or Americans. They feel protected by membership in group, are committed to their individual identity. They can, therefore, in academic meetings

The front you present to others is very important and you must be sure to present the right one. It is permissible to make mistakes as long as nobody important knows about them. Never show your incompetence in anything.

make such statements as "I know you all think I am a fool and that you will spit on my ideas, but I am going to tell you about them anyway." No American would dare to be so provocative.

The result is frequently a certain amount of skepticism about others. Salesmanship is taken for granted, while the need to establish close relationships— as with the French—does not necessarily hold in business. Germans do make friends and very close friends once the outer barrier has been let down. Americans let people in, but protect a hard core near the center so that people have the feeling they never know us, i.e., that we are all *image*.

Personal Relationships

Polychronic time brings people together and accentuates highly personalized relationships. But due to their need for privacy, the French protect themselves by not putting their name on the door. In their system you are either in their circle or you are out, and if you are out, they may not be too happy to see you. It is quite natural then that given this set of circumstances and the obstacles to be overcome, the French salesmen would "own" their customers, who are not the property of the company. It sometimes takes years to develop a relationship, which is why you have to work on several at a time. This is also why the French take their customers with them when they move.

Compartmentalizing M-time seals people off from each other so that personal relationships tend to be defined in terms of the job. Great care is taken to protect the privacy of *others*, whereas the French are preoccupied with protecting their *own* privacy.

Much of interpersonal behavior centers around communication. German communication, being generally low context, places great stress on words and technical signs, which is one reason they have to go into so much detail and why symbols of authority carry so much weight.

Interpersonal communication in
France depends more on the
high context messages of the
body and the face (movie actor
Fernandel is an example). High
context messages take longer to
learn to read accurately but are
much faster once learned—and
more dependable and trust-
worthy.

Propaganda and Advertising

In general, high context cultures
are more resistant to propaganda
and advertising, which must be
amusing and punchy—not serious—if it is to be effective.

Low context cultures are, in
general, quite vulnerable to
propaganda and to advertising.
That is, until they learn that the
agent behind the message is
untrustworthy, and then they
may mistrust all advertising.

Words from the man at the top
are, however, taken very seriously, even if one does not agree
with them.

People pay attention depending
on who is talking and how
forceful and convincing a case
they are making, regardless of
where they are in the organization. The Germans are tuned
and therefore vulnerable to low
context communications, regardless of source. Many are aware
of this, which is why they attach
so much importance to where
the source is on the scale of
political philosophies.

The Role of the Press and the Media

In high context centralized cultures such as France the press
seems to speak from a defined
center of power. Which center,
depends upon whether people

One might assume that in a
low context, closed score,
M-time culture the press might
be controlled. As a matter of
fact, the press in Germany and

listen or not as well as whether they are in sympathy with the point of view or not. the United States is one of the few sources of large-scale feedback on important issues. Without this safeguard, either country would be in a much more precarious position because of the propensity to "sweep things under the rug," to let people solve their own problems, not to fire incompetents, and the vulnerability of the system to strong personalities at any level. The press and the public media are the only ones who are freely allowed "to make waves." Of course, the responsibility is awesome, and is not always respected by members of the press.

Given two systems that are as different in their basic structure as are those of France and Germany, it is no wonder that rapprochement is often difficult to achieve. If one is advising people in the conduct of international affairs on either the governmental or the business level, I would suggest very careful selection of personnel, looking for those who are intuitive, sensitive, and superintelligent. Success in a cross-cultural situation requires much more talent than climbing the ladder of success in one's own culture. There are exceptions, of course (certain personality types sometimes find cultures that are vulnerable to their ministrations and wiles, and therefore do well even though they may not be unusually gifted). There is no doubt in my mind that to sell products in France involves very different rules than in Germany and takes more time, even though the German system may be more cumbersome.

PART II

Time as Experience

8 Experiencing Time

Since the beginning, mankind has been submerged in a sea of time. The sea is characterized by many diverse currents and countercurrents, fed by rivers from different lands. The rivers alter the mix and produce a unique chemistry of time for each location. Human beings, like fish in water, have only slowly made themselves aware of the time-sea in which they live. Like many important patterns in life, awareness of time is at first difficult to demonstrate. It is worthwhile to reflect a moment on the great differences resulting from such insights on the part of our forebears. When this happens, something really new is added, the first indications of which were present in Neanderthal burials in Europe somewhere between 70,000 and 35,000 years ago. Following the Neanderthals, Cro-Magnon hunters inhabited Ice Age caves in southern France and northern Spain, beginning about 37,000 years ago.[1] The Cro-Magnons also buried their dead. The cave deposits have also produced evidence that these men and women, who were the first modern human beings, had begun to make and record systematic observations of the phases of the moon, migration of game animals, the spawning of salmon, and possibly even the position of the sun at different times

of the year. Being able to record and predict such events as
the ripening of berries, fruits, and grasses, as well as the migration periods of different birds, fish, and game animals, greatly
enhanced the potential for survival of these early human beings
and made it possible for them to plan for the first time in human
history.

All this is known principally because of the efforts of one man,
Harvard University's Alexander Marschack,[2] an archaeologist who
made highly detailed examinations of Stone Age sequences of
scratches that appeared on the surfaces of bone tools and the
ribs of bison found in caves. Under the microscope, instead of
random scratches he found purposefully engraved marks that
proved to be unique. Each mark had been made at a different
time and with a different instrument! Marschack's evidence is
unequivocal. We have here mute testimony of the modest beginnings of observations and studies that were to hold the human
species spellbound for all time to come.

Much later, beginning sometime in the Bronze Age—two to
three thousand years ago—primitive but accurate models for
recording the movements of the sun, moon, and planets began
to appear all over the world. These "observatories"—some call
them computers—of which Stonehenge is the best known, made
it possible not only to accurately set the dates of religious ceremonies, but also to predict eclipses of the sun and moon. The
ability to predict seasons was absolutely essential as an aid in
planting and plotting the life patterns of the planet: predicting
when deer were in rut, when big game and birds migrated, when
the last frost could be expected to occur, when there were apt
to be storms. In fact, everything that impinged on our species
had some time period associated with it. These observations
were also of paramount importance in maintaining the proper
synchrony between religious ceremonies and festivals and the
seasons. At first, this knowledge was esoteric and held in the
hands of a precious few who controlled knowledge and kept the
secrets. During these early periods, time centered in the universe and in nature. The units were large. A day or half-day
(before or after midday) were the smallest units of time. The
week was unknown, months were merely a succession, and the
knowledge of the exact day on which the winter solstice took

place was in most societies restricted to one or two men, or to a small priesthood in the larger societies.

Following the Bronze Age and its massive computer-like formations such as Stonehenge, clocks apparently evolved from elaborate astrolabes, working models of the solar system used in astrology. Clocks did not appear in Europe until the fourteenth century, and were at first only owned by royalty and the very wealthy. By the mid-sixteenth century, well-established clockmakers' guilds had grown in Europe, while clocks were beginning to be sold in city markets.

The AE love affair with various kinds of timekeeping devices even exceeds the AE love of the automobile and in some ways may have made a deeper impact on our lives—albeit more subtly. It is the clock that is primarily responsible for our preoccupation with variable time (time dragging and time flying), which is the central theme of this chapter. It was the clock that provided an external standard against which to judge the passage of time—to determine whether it was "racing" or "crawling." Until then, people's internal clocks moved fast and slow, and usually in unison, so that few possessed an awareness of the speed at which time was passing. Even today, it is the presence of clocks that makes us aware of the passage of time. We know this from multiple experiences with peoples all over the world who do not or did not have clocks. As recently as fifty years ago, the Navajo Indians with whom I worked did not own clocks and had no need for them.

Before going into the complexities of variable time, it is necessary to say something about *extensions*,[3] of which clocks, watches, and calendars are examples. Extensions are basically tools and instruments, including tools of communication such as language. They are a natural product of most, if not all, living substance, although the extension process has been tremendously enhanced by humans. Examples among the less-evolved life forms are spider webs, bird nests, and territorial markers. Mankind has evolved its extensions to such a point that they are beginning to take over the world and may ultimately make life impossible unless they are better understood.[4]

Extensions are remarkable because they can be evolved at virtually any speed, whereas life itself is the product of small accumulative changes in which the generation is the shortest interval in which genetic changes may occur. Therefore, small animals like flies, and bacteria and viruses that reproduce very fast, can evolve adaptations to the environment in very short time periods. Worldwide resistance to DDT by flies and mosquitoes is one of the best-known examples of this sort of adaptation.

If human beings had had to evolve culture genetically, it is doubtful if we would have progressed beyond the Stone Age. To speed up evolution and achieve flexibility in meeting environmental challenges, humankind began to evolve its extensions. The human species, however, paid a price when it chose the extension route. Extensions are a particular kind of tool that not only speed up work and make it easier but also separate people from their work. Extensions are a special kind of amplifier, and in the process of amplification, important details are frequently left out. What gets left out is largely a matter of chance and sometimes what is left out may be more important than what is amplified.

One of the most important central issues to be understood about extensions is that they are rooted in specific biological and physiological functions. They originate in us! Properly read, one can tell an incredible amount about human beings by studying their extensions. In fact, there is little that can't be discovered. Extensions can be viewed as externalized manifestations of human drives, needs, and knowledge, and they even reflect our unconscious drives. Given the current state of the world, this is sometimes difficult to envision, but no one else made our extensions—we did. Examples of extensions are: the telephone extending the human voice, television extending both the eye and the ear, cranes extending the hand and the arm and the back, computers extending the memory and some of the arithmetic parts of the central nervous system, telescopes and microscopes extending the lens of the eye, cameras extending the visual memory, knives extending the cutting and biting capabilities of the teeth and fingernails, and automobiles extending our legs and feet.

There is one more point to be made, and that is that whenever something is extended, the extension begins to take on a life of its own and quickly becomes confused with the reality it replaces. Language is an excellent example. The process was best described by Count Alfred Korzybski when he formulated the principles of general semantics.[5] Korzybski stressed that the word is not the thing, it is only a symbol. This is one of the most difficult concepts for humans to grasp. It would seem that human beings must learn over and over again that the map is not the terrain.

In another work, I formulated the principle of extension transference, which holds that any extension not only can but usually does eventually take the place of the process which has been extended.[6] This principle is illustrated by the way in which we have taken our own biological clocks, moved them outside ourselves, and then treated the extensions as though they represented the only reality. In fact, it is the tension between the internal clocks and the clock on the wall that causes so much of the stress in today's world. We have now constructed an entire complex system of schedules, manners, and expectations to which we are trying to adjust ourselves, when, in reality, it should be the other way around. The culprit is extension transference. Because of extension transference, the schedule is the reality and people and their needs are not considered.

Time Passing and Time Dragging

Time "drags" when the body clock and the clock on the wall are out of sync. Time dragging is a synonym for not having a good time. The message that time is dragging can be used to alert individuals to find out what it is that makes them feel that way. Recognizing these little cues—like time dragging—is important, because it is becoming increasingly clear that our unconscious is where the organizing, synthesizing core of our personality is located. Many, if not all of us, attempt to reduce alienation and try to bring the conscious part of ourselves in line with our unconscious. The gap between the unconscious and the conscious is not inconsiderable. After a certain point, when this gap is too wide, people's lives are diminished. The strain of trying to bring the two parts together makes them less productive and less happy. A sense that time is dragging should be a cue to take a closer look at the state of one's psyche.

Maggie Scarf in her book about female depression, *Unfinished Business,* states that depression has a considerable biochemical component. That is, it can be treated with drugs or with a combination of drugs and psychotherapy. It doesn't matter to the individuals suffering from depression whether their troubles are due to an imbalance of the chemistry of the body or whether they are of psychogenetic origin. The hurt, the suffering, and the debilitating paralysis are the same. What makes depression doubly unbearable is that time oozes at a snail's pace. One of Scarf's subjects, Diana, a woman who had just tried to kill herself, speaks of the "molasses-like feeling of being stuck in endless time" (p. 347). Scarf also states that "Menopausal depression is a biological time bomb that can explode in those years in a woman's life . . . when fertility is . . . ending."

As a young man working on Indian reservations, I often saw Navajos and Hopis patiently waiting around trading posts at the agency in Keams Canyon, Arizona, or at the hospitals in Keams and in Winslow. I realized that it was not possible to imagine myself in their shoes. There was a different quality to the Indian's waiting from my own. In this respect I was no different from other white men. We were all impatient, always looking at our watches or the clock on the wall, muttering or fidgeting. Yet an Indian might come into the agency in the morning and still be sitting patiently outside the superintendent's office in the afternoon. Nothing in his bearing or demeanor would change in the intervening hours. How could that be? As a child I had encountered the same phenomenon in Indian pueblos in northern New Mexico, in the houses of Indian friends, and in the towns of Santa Fe and Taos. The Indians would exchange visits with my family, as well as with many of the artists in the area. As it happened, white men's hours and days for visiting never seemed to mesh with those of our Indian friends, with the result that whoever did the visiting would find themselves waiting. A strain gauge attached to the bottom of a bench would have recorded the difference in waiting behavior quite graphically. We whites squirmed, got up, sat down, went outside and looked toward the fields where our friends were working, yawned and stretched our legs, and made innumerable other signs of impatience. When the roles were reversed, the Indians simply sat there, occasionally passing a word to one another.

Later, as a grown man working and visiting in other countries, I encountered this same difference. It was quite evident that my time was not their time. Arab men who spend hours on end— in fact, all day—talking to their friends in coffeehouses still amaze me. Even people in Paris cafés exuded a different air from what I had experienced at home. In Paris the same people could be seen sitting day after day watching the world go by. The restaurateurs were tolerant of Bohemians. Everyone knew that artists didn't have much money and therefore couldn't afford to heat their studios, so they would sit in cafés and soak up the heat. Furthermore, the experience of time varies in detail from class to class, by occupation, and sex and age within our own culture. Did you ever notice how impatient young children are in our culture? "Mommy, how much longer are we going to have to wait? I'm getting tired." One would never hear a whimper from Indian children. Occasionally there was an almost imperceptible tug and the mother would reposition the child or uncover a breast so her child could nurse. The whole process would happen with no break in rhythm and so naturally that I almost missed it. Clearly culture patterns such as these must begin very early in life and be in place at the time of birth.

In the United States and in AE cultures generally, people seem to assume that time is a given (as it was in the Newtonian sense), that it is the same wherever one goes in the world. Clearly this is not true, but in order to gain insight into the time systems of others, we must know more about our own. How does one gain such knowledge?

What Literature Can Teach Us

I have grown to depend on literature as a source of discovering people's preoccupations. The novelist and the poet reflect the principal preoccupations of people and their times.[7] Henri Bergson was obsessed with time and considered it an enemy. Proust, like his fellow countrymen, was preoccupied with time and felt that time and memory were inseparable. What is fascinating

about these two men is how characteristic of AE culture they
are.

Clearly, the novelist must come to grips with time, and how he
or she handles it is a good index to mastery of his craft. James
Joyce sees us imprisoned by the "narrow confines of linear time."
Joyce's protagonist Stephen Dedalus thought it was impossible
to separate the clock from the experience of the viewer. And in
a way he was right. For Bergson, "becoming" was the essence
of time. All these writers are conscious of being conscious.
Hurdling the barriers of language, they land right in the middle
of time. Time to them was the equivalent, in fact the quintes-
sence, of consciousness. What most of these writers really did
was use time as a tool to pin down consciousness.

Time is, of course, a major device in the works of Virginia
Woolf, Aldous Huxley, Franz Kafka, Thomas Mann, Thomas
Wolfe, and William Faulkner, to mention only a few writers.
Clock time and mind time as two distinct and separate forms
are recognized by all of them. Bergson saw duration as the
meaning of life itself, while Kafka made inner time real. Yet
Kafka annihilates time by turning reality into a dream, which
is what gives his work its surreal quality.

All of these authors implicitly and explicitly accept duality
as axiomatic in nature; individual and universal, will and idea,
concrete and abstract, artistic and materialistic, separation and
merging, present and past, past and future, present and future,
outside looking in and inside looking out, life and art, time and
eternity, sympathy and detachment, mysticism vs. humanism,
instantaneity and eternity, and symbolic and allegorical. Yet duality is nothing more or less than the way in which AE cultures
categorize virtually everything. Physics and anthropology tell
quite a different story. But the reader should know that duality
is, as Einstein put it, something which one "imbibes with one's
mother's milk." We all come by it naturally, which diverts our
attention from multiple causation. Stimulus response expresses
the duality pattern, cause and effect. The reader should know
that duality comes naturally if one is an American of North
European heritage and that it will be less natural for him or her
to look for multiple causes than for people brought up in cultures that take a pluralistic view.

Time Compression and Time Expansion

Time compression and time expansion are two subjects of continuing fascination for AE peoples. Time compresses when it speeds up. This is evident in emergency situations when one thinks one is about to die ("My whole life flashed before my eyes") or when there is extreme pressure in order to survive. An example would be the case of Major Russ Stromberg, Navy test pilot, testing the Carrier AV-8C.[8] Stromberg had just been catapulted from the deck of the aircraft carrier *Tarawa* and he realized that his plane was not developing power. This eight-second scenario of how he dealt with the emergency and survived took forty-five minutes to describe. "I was very surprised by the whole evolution of the thing. Everything went into slow motion. *After about one second*, about seventy-five feet after I started rolling, I knew I was in deep trouble" (italics added). First, Stromberg had to see if the engine could be brought up to power by switching off mechanisms limiting takeoff power. That didn't work. There was no way to get the engine up to power in the five seconds remaining before the plane would hit the water at over a hundred miles an hour and disintegrate. Ejection was the second option. However, to eject at the wrong moment would also have meant certain death. Even with only two or three seconds, he had the time to look around so that he could pull the ejection handle at just the right moment: thirty feet above the water. Stromberg ejected and fortunately avoided the crash site by only a few feet. This meager description cannot possibly cover all the possible alternatives to decisions that Stromberg ultimately had to make—at the right time, in the right order, and without panic. If he had been on normal time, none of this would have been possible. If that capacity to expand time—in this instance to about 300 percent of normal time—had not been built into the human species, it is doubtful that the human race would have survived.

I once had a similar—although not as dangerous—experience when I found myself involuntarily sharing, with a mountain lion, quarters from which there was no exit. I had had some dealings with a naturalist of sorts who had cheated me. To distract me from trying to collect my debt and suspecting that

I might be a sucker for animals (I had shown great interest in a chipmunk he carried in his pocket), he asked if I would like to see his new mountain lion. While in the process of looking at the lion, I discovered to my horror that this man had inadvertently (?) left the door of the cage unlatched so that the mountain lion escaped into the narrow passageway where the two of us were standing. My first awareness of what had happened was when I felt something brush by the calf of my right leg. Then as I watched the lion lick a spot of grease next to my toe, time slowed down. The experiences of others with wild animals flashed through my mind: "If you are afraid, the animal will sense it and kill you." I was not afraid; that came later when I was safe again. But what to do? How to get out of this mess? Putting years of experience with animals to work, while I mentally alternately reviewed and rejected a half dozen options and their scenarios, the only workable solution seemed to be to make friends. This mountain lion, I discovered, could be approached, and it even purred like a model-T Ford. Having assured myself that I wasn't going to die just yet, I didn't push the relationship, either. Eventually the mountain lion (his name was Jim) was back in his cage and I departed, having completely forgotten my reason for being there in the first place. How many times events of this sort have happened in the history of the species of course is not known, but having lived an active, outdoor ranch life in the wilds of New Mexico and Arizona as a young man, I do know that emergencies are legion and that this capacity to slow down time in an emergency has saved me more than once.

For city dwellers, living a life cut off from nature, swaddled in technology and creature comforts, it is difficult if not impossible to visualize what it would have been like to live the life of our forebears. As anyone who has spent much time in the open knows, a built-in, variable time sensor is necessary for survival. Perhaps it is still necessary today to confront a new set of dangers, those of urban life.

Concentration and Time Perception

The degree of concentration required to complete a task is related to how fast time is perceived as passing. The events that cause a particular person or group to concentrate so hard that they "lose all sense of time" can be attributed to multiple causes. Top athletes are known for being able to concentrate, and frequently when they fail to do as well as they hoped, they put the blame for their failure on a slackening of concentration.

Concentration of any sort obliterates time. Some of the most impressive and well-documented examples of this come from the new field of microsurgery.[9] Microsurgeons work with a microscope to reattach parts of the body: arms, legs, fingers, hands, toes, and eyes. Microsurgeons work as part of a team— teamwork not only makes these incredibly difficult, demanding operations possible, but also seems to have a supporting effect on the surgeon, giving him energy and assurance when his vital forces are running low. In one operation, the team literally worked around the clock (twenty-four hours and twenty minutes) to reattach the four fingers of an eighteen-year-old girl who had caught her hand in a printing press. This meant sewing together nerve ends, muscles, tendons, blood vessels, and skin, and working out procedures so that the fractured bones could knit. The doctor said, "I wasn't conscious of time." It is also interesting to note that surgeons of this type have to keep themselves in superb physical condition and avoid caffeine for twenty-four hours before the operation, and none of them smokes!

One does not need to be a surgeon to concentrate; almost anything that is sufficiently involving will suffice to make a normal person concentrate. A talented young filmmaker from Switzerland experiences the time between meals when she is working as mere minutes. Suddenly she will feel hungry and realize hours have gone by. Only when other people are involved is she conscious of time.

Imagery and Time

Another instance of time perception which is virtually impossible to replicate in the laboratory is what went on inside Mozart's head when he was composing music. It is only possible to look at his extraordinarily productive capacity and take his word for what he experienced when he composed. To put this discussion of Mozart in context, one has to know that there are two ways of doing virtually anything creative.[10] As children, most of us learned early that there are some persons who can do their arithmetic in their heads, while others solve the same problems outside their bodies with the aid of pencil and paper, or chalk and blackboard. This same process can be observed in any field: choreography, architecture, industrial design, sculpture, painting, writing and composing, even skiing and dancing. Teachers happen to prefer the second method because they can see what is going on and correct "mistakes." It makes them feel useful and in control. The first method, however, is faster and more creative.

Both Mozart and Beethoven could compose in their heads. Beethoven could compose for the strings, listen to the results, and then feed in the brasses to see how it sounded. But insofar as I have been able to learn, he did this sequentially; that is, he listened to his own music just as we do when it is played in a symphony hall by a live orchestra, with head time and real time reasonably in sync. Mozart was different. Something about the way in which his central nervous system was organized enabled him to experience his music all at once. It is possible, of course, that Beethoven was a left brain genius, and Mozart a right brain holistic composer. But intuition leads me to believe there was more to it than that. The brain is organized very much like a hologram.[11] That is, the information is stored everywhere at once so that it is impossible to fix a memory in a specific location in the brain.[12] It also seems to be stored in layers, so that an individual who has a stroke may lose one language but retain another. Mozart's experience provides us with a small clue as to how certain individuals frequently have the capacity to span time—in effect to see into the future. In large groups, for example, once music starts, there will be those who will know what's coming, because like Mozart they experience what is going on in the present as a portion of a unified entity that is played out in a sequential manner. Clearly, when Mozart was composing, the experience of time must have been a totally different process than it was for his colleague Beethoven.

Howard Gardner, a psychologist writing about the two men in *Psychology Today*, could not bring himself to believe that

Mozart could experience something as complex as a symphony all at once. Beethoven apparently could hear music in his head, but he did not hear the details, which Mozart did. Beethoven had to translate his music onto the page, and then work to get the written score to conform to the mental image (just as Einstein had to translate his visual and physical images into words and then into mathematics). Mozart, however, apparently not only had an incredible creative capability, but was also a simultaneous translator. Remember that all notation systems and all extensions by their very nature leave some things out, so Beethoven's difficulty in writing down his music was in part traceable to the fact that the written music seldom achieves the perfection and congruence of what one hears in the head. Beethoven was famous for working over his manuscripts. Being unusually sensitive to congruence, he would work and work until he found just the right note.

The difference between creating inside oneself and creating outside by means of an extension is basic and crucial. The two processes are entirely different. It takes ten to fifty times longer to do things outside the body than inside. Several designs can be considered and rejected in the same time that one is putting ideas on paper. Extensions speed up change, but slow down productivity, particularly of an integrative, complex type. The other essential distinction is between experiencing things sequentially or as discrete units. Again, the external sequential mode is much slower. The artist or scientist who sees a complex form all at once will have fewer problems externalizing or translating into symbols than the individual who has to tease his product out in bits and pieces, externalizing something virtually without form from his unconscious which he then assembles outside his body on paper, canvas, clay, or a dance floor.

Age and Time

Age affects how people experience time. The observations on this are well known, so it is only necessary to outline briefly what has been the experience of everyone I have ever talked to or read about: the years go faster as one grows older. At the age of four or six, a year seems interminable; at sixty, the years begin to blend and are frequently hard to separate from each other because they move so fast! There are, of course, a number of common-sense explanations for this sort of thing. If you have only lived five years, a year represents 20 percent of your life; if you have lived fifty years, that same year represents only 2 percent of your life, and since lives are lived as wholes, this logarithmic element would make it difficult to maintain the same perspective on the experience of a year's passage throughout a lifetime. There is a cultural factor as well. In cultures like our own where the group past becomes dim as it recedes and where something that happened twenty years ago is considered "ancient history," the total effect leads to a deep impression of time speeding up. The more that one has buried in the past, the faster the present will appear to move. This is in contrast with cultures where the past is kept alive, as it is in the Near East, where virtually everything in today's world is seen as having been rooted in the past.[13]

Time in Relation to the Size of the Job Ahead

There is also the matter of how much is stored in the memory in relation to new things learned. A twelve-year-old or even a graduate student looking at all there is to be learned has a very different view than someone who has had a Ph.D. for twenty-five years. I had an opportunity to test this once in my late fifties when I learned to fly. The amount that one has to learn to perform in a creditable and safe manner is on the order of that required when taking an advanced degree. The difference is that in a plane it is difficult to go much more than a few minutes or even seconds without being tested. During landings, time compresses. There you are, up in the air with the landing field on your left. Having already contacted the control tower several miles out and entered the traffic pattern, you must perform a wide range of tasks with skill before you are safely on the ground again. You must slow down to the proper speed, maintain your designated altitude until you make your next to last turn toward the field, keep your engine at the right speed and temperature, turn on carburetor heat if the engine is not fuel injected, lower your landing gear, set your flaps, and when on final approach, maintain a uniform rate of descent while keeping the proper angle of descent so that you land neither long nor short, maintain a safe flying speed so you won't stall, all the time maintaining contact with the tower, watching for other planes (they are always there), eyes on the wind sock for wind shifts and, in the Southwest, for dust devils. All these things require constant adjustment so that the entire process unfolds in a coordinated, integrated, rhythmic manner. As you watch the numbers at the end of the runway get larger and larger you mentally remind yourself that, while your approach has been good so far, you haven't even landed the plane yet and that in a few seconds you are going to have to shift gears and begin a whole new series of maneuvers in which things will start happening very fast and where there is little leeway for error. For me, learning to fly was, in many ways, like being a child again. After I had learned enough to know what I didn't know, time slowed down and I was overwhelmed by a feeling that I would never master the complex interrelationships necessary to be both skilled and safe in the air. In due course I mastered what had to be mastered and was duly licensed. In the process I also learned a great many other things, such as the fact that the experience of time is tied in some mysterious way to the perceived size of the job ahead. As one works one's way from the outside into the select membership of a group one aspires to join, one's perception of time changes. Ask any academician how he experienced the time before he was tenured and after he was tenured.

Piaget

Jean Piaget was one of Europe's most gifted and innovative intellectuals—a giant of a man, who never lost interest in learning, nor the "scientist's capacity for surprise."[14] There are so many things he was right about, and he added so much to our knowledge about child development that one feels almost guilty criticizing even a small portion of his work. Piaget had a great capacity for putting his finger on the very thing that made a difference. Nevertheless, like all of us, Piaget was a product of his times and the preoccupations of those times. Not being well acquainted with the basics of primary level culture, Piaget assumed that "logical thought" was natural—that the potential for logical thought was built-in and emerged as a function of maturation. Piaget was mesmerized with the process of logic as it "emerged" in the developing child. His analysis[15] begins with a discussion of Newton, Einstein, Descartes, Kant, and other leaders in the Western philosophical and scientific tradition and the world that they have created. Piaget then proceeds to develop some terribly ingenious ways of testing rather precisely just where a child is in the mastery of this world—including its underlying assumptions, such as, that perception is a logical process (which most definitely it is not). There is a nod in the direction of informal time and some recognition that duration is not necessarily perceived as a constant by the child. However, there is not even a hint that the entire perceptual process is not only learned and modified by culture but is constantly influenced by context—that the perceived world is a transaction![16] What Piaget's work does tell us is a great deal about acculturation in the West and the principal preoccupations of our own culture. His generalizations should not, however, be applied to the development of children in other parts of the world. When Piaget studied the child learning the basics of time and space, he did not realize that the child was also learning our own system of logic. There must be hundreds of different systems of logic in the world, some of them high context, some of them low, some in between. Most of them are learned. Like all extensions, these logic systems leave things out. Our system in its most developed form leaves out context, and that is a very significant omission, indeed.

Piaget has much to say about the child's perception of time.[17] Nevertheless, Piaget's methodology, though satisfying AE demands for linearity, is still culture-bound. I say culture-bound because the child has not yet been taught to distrust his senses, nor has he learned that one of the things that one must deal with as an adult in the Western world is to expect to be deceived by containers of different sizes and shapes. While Piaget quite correctly identifies time with space,[18] his failure to distinguish physical time (see chapter 1) from core culture time, from profane time, does not provide a clear picture of how the child is either experiencing or structuring time at different stages in his development. In *The Silent Language,* which dealt with informal time—the time which is not explicit—I reported on how children attempt to learn the meaning of terms such as "a while" and "later." Both of these are highly dependent on context. A particular mother means one thing when she says, "I will only be gone a little while, dear," whereas another mother, next door, means something quite different when she utters these same words. This is the sort of thing children have to cope with when growing up, which has little to do with philosophical discussions of the "meaning of time." But what the child really wants to know is how much perceived time will transpire before his mother gets back and how he is going to feel about that period when he is alone. Attempts on the part of mothers to allay their own guilt and anxiety usually don't tell him what he needs to know. They do tell him, however, that there is a kind of time that is measured with a "rubber yardstick" and the length of that stick is a function of events over which he has little or no control. Mood and psychological states have an incredible effect upon the experience of the passage of time. This may explain part of the child's difficulty in learning the informal time system. When the mother is away, time drags, because he only feels reassured, happy, and safe when the mother is nearby.

Mood and Time

Almost from the beginning, there has been great interest as well as considerable speculation among AE peoples concerning the effect of mood on how people estimate time—is it flying or crawling? Is there anyone who has not had the experience of being deeply involved in a happy transaction and suddenly realized: "Oh, my goodness! I was having such a good time. I had no idea it was getting so late"?

Since remarks of this nature would seldom be made in a polychronic culture such as Lebanon or Syria, the whole matter of mood as a factor in the perception of time must be looked at anew. It works that way in our culture, but in how many others? Sometimes mood is a secondary factor in the psychological environment. Other factors, such as depression, which has already been mentioned, or even body temperature, may be determinants. Professor Hudson Hoagland, a famous physiologist who taught for years at Clark University, discovered that body temperature affects the perception of time—a fact that is not surprising if one thinks about it. The higher temperatures speed up the body clocks along with everything else. Lower temperatures slow them down. Diabetics, for example, frequently have subnormal temperatures, which would make time appear to move faster. Liquor works on two levels: because of the sense of stimulation and conviviality, time initially moves faster, but since liquor is a depressant, the ultimate effect is for time to drag. Hangovers can seem eternal!

Anniversaries

Anniversaries are a cyclic manifestation of time. Yet, because of the subtle and ubiquitous nature of our own internal timing mechanisms, many of us react to anniversaries of which we are not even aware. A logic of a different type is at work here. This writer has been afflicted with a springtime depression for years. This depression was a mystery until he finally realized that his parents had been divorced in the spring and that the world had collapsed when his mother, whom he dearly loved, left the family. One reason he eventually was able to sort this out was because he lived in many places where there were different associations with spring. That lingering depression, however, had to be dealt with independent of the surroundings. Some internal clock kept tabs on time alone. By the time people reach the age of sixty, there are so many of these hidden anniversaries— anniversaries of triumph or of failure and disappointment—we sometimes don't know whether it is an immediate situation or something we have forgotten in the past that causes us to feel cheerful or depressed, or is causing time to speed up or slow down.

Emotions, Psychic State, and Time

Time away from a loved one moves at a snail's pace, while a rendezvous is over before you know it. Not too much is known technically about the effect of the different emotions on the passage of time. One could assume that, in general, the positive emotions would cause time to speed up, while the negative ones would make it crawl. But perhaps not. Since many of the hormones associated with emotions have been identified and are available for experimental purposes, it should now be possible to develop double-blind experiments on the effects of these hormones on perceived time.

In the last few years, an increasing number of Americans have been meditating and the interest in meditation has been very great.[19] A number of psychologists interested in the effects of meditation on people have developed tests trying to discover what meditation is all about—the physiological, psychological, and neurological accompaniments of meditation. This voluminous research has demonstrated that meditation alters the heartbeat, blood pressure, respiration rate, galvanic skin response (GSR), and brain waves. Meditation is also a great antidote for stress. Since one of the leading causes of stress in the Western world is associated with the pressures of time, one would expect to find that meditation alters perceived time. These expectations are well founded. In an interesting, insightful article on the subject, Keith Floyd[20] describes situations in which time first slows down then stops altogether. Awareness of time ceases when the meditative state is very deep. Apparently, this is a different state from that of the artist who is working in his studio or the scientist in the lab who "loses track" of time. It would be interesting to know whether these two really are different. One approach would be to select scientists and artists who also meditate and have done so for a long-enough period to have gained the proper level of proficiency and simply ask them to describe how the two states are alike or different. Another way would be to replicate the brain wave respiration and heart rate studies conducted earlier on meditating subjects, but this time concentrating on scientists and artists. It would be necessary to use telemetering instruments, since no scientist or artist could work with all those wires and recording instruments immobilizing him.

One is compelled to assume that these different states are functionally advantageous in humans and may be one of the

most important of the many feedback mechanisms that tell people
how they are doing. It is extraordinary and paradoxical that the
very activities that are most rewarding and satisfying are those
in which time is experienced as passing with extreme rapidity
or in which the sense of time has been lost completely. Being
able to accomplish all the things that must be done to save one's
life in an emergency is a far cry from annihilating time during
meditation. One has the feeling that here is an instrument which
man has used much too casually, and about which not nearly
enough is known.

The Perception of Time as Mediated by Space

Since 1976, a psychologist and professor of architecture at the
University of Tennessee, Alton De Long, has been conducting
precise observations under carefully controlled conditions on how
different people experience the passage of time when they are
interacting with environments of different scales. De Long demonstrates once and for all that time and space are functionally interrelated. The perception of time is not only influenced by the
multiple factors mentioned earlier, but by the scale of the environment as well.

The process is truly remarkable and, among other things,
illustrates the inherent logic of the central nervous system.[21]
Apparently, this type of logic is something all normal human
beings have. It is altered on the conscious level by culture, but
retained on the unconscious level. Whether or not we are aware
of the process, if one part of the system is changed, the central
nervous system tries to make accommodating alterations.[22] I have
discussed the capacity of the human central nervous system to
keep things in balance in an earlier work.[23] This principle applies
even when the person suffers from psychosis. When the schizophrenic suffers from perceptual distortions of body boundaries,
the human brain will still try to maintain a logical balance.
Anyone with a perceptual distortion of this sort is, of course, under a dreadful strain. Everything changes—all relationships to
the physical world, and to people as well—simply by virtue of
this single aberration in perception. Discussing this with a psychiatrist colleague, Dr. Harold Searles, I said, "These patients

of yours who experience their body boundaries as expanding should—if my previous observations are correct—experience that they are light, like a balloon." Searles then told me that on that very day a patient had brought him a crayon drawing of an object that looked like a blimp with short appendages. It was a self-portrait! The patient had drawn in lead shoes on her feet to keep from being blown away because she felt so light.[24]

It would be expected, therefore, that the brain could and would make comparable compensating adjustments in time, and that is actually what happens. Under proper conditions, subjects will increase interaction rates in an environment to stay in agreement with the scale of that environment. An environment reduced to $\frac{1}{6}$ of normal size can actually program the central nervous system in such a way that subjects who project themselves into that environment will hold their own internal time perception constant. This adjustment process results in a compensating speedup in the processing of information by a factor of six. What is experienced as one hour's work in the model is actually only ten minutes by the clock. Using a 1:12 scale, the experience of an hour's work takes five minutes of "real time."[25] Furthermore, EEG (brain-wave) studies conducted as part of De Long's research seem to indicate that the mediating mechanism is the brain itself. The brain speeds up in direct proportion to environmental scale.[26] It should be noted that beyond 1:12, environmental effects begin to fall off and the test environment is simply coded differently in the brain. Why the break-off point is at 1:12 instead of 1:20 or 1:50 we do not know. This ratio is, however, apparently one of the basic ratios for the average AE individual. People who can work at greater ratios must have an unusual and unexplored advantage over the rest of us.

How did De Long set up his experiments to get such remarkable results? The procedure he used was somewhat like playing with a furnished dollhouse. Those who have done so as children will remember the time warp that occurred when they were really deep into their play. De Long's environments were selected for four different scales: $\frac{1}{24}$, $\frac{1}{12}$, $\frac{1}{6}$, and full scale.[27]

Subjects were given masks that screened off peripheral views of full-size objects and asked to project themselves into the environment by identifying with one of the human figures that had been placed there by the experimenters. Subjects were not allowed to touch the figures but were told to participate in some imaginary activity of their own choosing and to indicate when thirty minutes had elapsed. The investigator told them when to start. The subjects signaled when they felt thirty minutes had passed. The experimenter timed the interval with a stopwatch. De Long took particular pains to be sure that his subjects did not think they were being tested on their ability to judge time. They were asked to be as subjective as possible (just as you would be if you had been waiting for the doctor for an hour). Their feelings were to determine at what point the thirty minutes had elapsed.[28]

What does this mean? Simply that, provided these studies can be sufficiently refined so that they can be replicated, for selected situations we should be able to look forward to a time when some kinds of decision-making tasks can be accomplished in $\frac{1}{6}$ to $\frac{1}{12}$ the normal time. After familiarizing people with the effect of environmental variables, it should be possible to give an individual up to 12 hours' experience in the course of an hour. De Long stresses, however, that people should not be subjected to miniaturized environments for periods in excess of those they work in normally. For someone used to working an 8-hour day, the maximum immersion in a $\frac{1}{12}$ scale environment would be 40 minutes. How many simulated (1:12) days people could work in such environments at such increased rates of speed is not known. My intuition tells me that the human species should use extreme caution in matters of this sort where so little is known and where the field is unexplored. Our interest in this chapter is, of course, the experience of time and the factors influencing that experience. De Long's study remains one of the few that relates the perception of the passage of time to the environment context under controlled conditions.

Estimating Real Time

The perception of time is certainly deeply embedded in context as well as dependent on situation. But what about the other side of the coin? We have been describing how people in different situations, moods, and conditions perceive time as speeding or dragging. How well do human beings do when it comes to judging time in real-life situations where performance is timed to the split second, such as downhill races with toboggans, skiing, and automobile racing? As it turns out, people do extraordinarily well. Within the past five years it has become generally known that humans can "image" time quite accurately in their heads, if they are asked to do so in specific, real-life situations, such as traversing a familiar downhill ski course. Coaches in all sorts of fields have begun to ask the members of their teams to time themselves with stopwatches while imagining a run over a particular course on which they will later be competing. Contestants try to image every part of the course—each curve, each soft spot, all of the straightaways—running the course in their head in exactly the same way that they would when actually covering the course in competition. Drivers and athletes not only come within seconds and hundredths of seconds of their actual performances, but also have been able to get valid practice in this way. It is safer, saves wear and tear, saves course and pit stop fees as well as fuel costs. Both coaches and participants feel that the practice obtained in this way is valid. Dancers and acrobats will recognize parallels with the kind of mental practice they are used to doing. It is extraordinary how accurate people can be in their timing of these simulated runs.[29]

How time is experienced is then a function of many things. It is situation- as well as culture-dependent. Once more I find myself being simply dumbfounded by the extraordinary variety as well as the tremendous range of capabilities of our species. However, humans are subject to getting stuck in a rut. Time ruts are no exception.

It might be helpful if there were wider understanding that in virtually any situation it is possible to be both positive and creative about one's life and that few things are more crucial to life than the use and experience of time. For as we shall see in chapters 9 and 10, it is the balance wheel of culture's synchrony which keeps us in phase with one another.

9 The Dance of Life

It can now be said with assurance that individuals are dominated in their behavior by complex hierarchies of interlocking rhythms. Furthermore, these same interlocking rhythms are comparable to fundamental themes in a symphonic score, a keystone in the interpersonal processes between mates, co-workers, and organizations of all types on the interpersonal level within as well as across cultural boundaries. I am convinced that it will ultimately be proved that almost every facet of human behavior is involved in the rhythmic process.

Since our topic is quite new, it is not surprising to discover that, unlike astronomers studying the universe or scientists searching for a cure for cancer, there are very few people involved in the study of rhythm.[1] Rhythm is, of course, the very essence of time, since equal intervals of time define a sequence of events as rhythmic. In the sense that rhythm is used here, it includes much more than the productions of musicians and dancers, although they are part of this process too.

First, let us begin by thinking small. Almost thirty years ago, when I seriously began studying proxemics (the use of space and man's spatial behavior),[2] it wasn't enough to simply observe that AE Americans did not like to be approached too closely during conversations and were, for the most part, averse to extensive touching or sensory involvement with people whom they did not know well. The fact that many Americans commented on their proxemic relations with Arabs and other Mediterranean peoples was interesting and relevant, but we needed to know more about what was actually happening. For example, how did people know when others were too close? What kind of measuring rods were they using? What was the physiologic-sensory base in which proxemic behavior was rooted? To answer these questions, a wide variety of observations and recording techniques were developed. One of the best, most effective, and reliable methods was cinematography.

Film after film of people interacting in normal situations was made. I filmed people in public spaces, in parks, on the streets, at festivals and fiestas, and in the laboratory under controlled conditions. Film provided us with not only a wealth of data to study but also a relatively permanent record to which we could refer time after time. There are many different methods for analyzing human interaction on film, as well as on video tape, but I will not attempt to describe the many techniques, because this is a technical matter for the specialist.[3]

Three things were apparent from the beginning in kinesic (the study of body motion) and proxemic research films: 1) Conversational distances were maintained with incredible accuracy (to tolerances as small as a fraction of an inch); 2) the process was rhythmic; and 3) human beings were locked together in a dance which functioned almost totally outside awareness. The out-of-awareness character of this behavior was particularly true of AE cultures and somewhat less true of African cultures, where the people are more conscious of the microdetails of human transactions.

Not only did we record that regular proxemic dance on film, but small experiments in the living laboratory had produced similar results. Experimentally I have backed people across a room, maneuvering them into corners by advancing a fraction of an inch at a time while we were conversing. My subjects were oblivious to the fact that they were adjusting their own conversational distance approximately every 30 seconds. To main-

tain a distance that was comfortable, they had to move. It didn't
seem to matter who the individual was, trained observer, scientist, businessman, or a clerk in a store. The sample included people of all descriptions and classes.

I discovered a system of behavior going on under our very
noses about which virtually nothing was known. It was known,
however, that people respond proxemically in all cultures. Whenever the proxemic patterns and mores were violated, people
reacted in readily observable and predictable ways.

If behavior of this sort could be identified through the study
of man's use of space, what might we expect to find in the study
of time? As a matter of fact, one finds behavior just as remarkable, possibly even more so, which parallels the results obtained
from proxemic studies. A person's structuring of his or her own
rhythm is an extraordinary process in which only a fraction of
the possible implications have yet been gleaned.

In 1968 I initiated a program of interethnic research in northern New Mexico,[4] where there is a mix of three cultures: the
Native American–Pueblo, the Spanish American, and the Anglo
American. Each maintains its own identity, but people meet, do
business, attend ceremonies and celebrations, make love and
fight, as well as mix in various proportions on the streets and in
public places like the plaza in Santa Fe. The dances performed
by the Pueblo Indians as public exhibitions of what in other
circumstances are sacred dramas are ideal for cinematographic
research. Everyone photographs everyone else, so one more
camera makes no difference. Having grown up in northern New
Mexico, I realized that I was already programmed to much of
what was being recorded on film. However, I was not prepared
for the richness and the detail of those visual records when they
were subjected to the frame-by-frame analysis of a time-motion
analyzer. Unfolding before my very eyes was a perpetual ballet.
Each culture, of course, was choreographed in its own way,
with its own beat, tempo, and rhythm. Beyond this there were
individual performances, pairs dancing out their own dramas,
and beneath all this was the truth of interpersonal encounters—
particularly those of the interethnic variety—the specifics of behavior that may engender misunderstanding, prejudice, and even
hate. Life unfolded in that step-by-step, frame-by-frame film

analysis. Events that occurred in fractions of seconds (too fast for people to notice and analyze under normal circumstances) could be seen and studied for the first time. Façades fell away and dissolved in front of my eyes.

This happened when I first began studying the interaction patterns of the three groups who inhabit the Southwest United States (AE whites, Spanish Americans, and Native Americans). To be certain that I wasn't just "seeing" things, I took the precaution of asking John Collier, Jr.—one of the most talented and insightful individuals in the field of visual presentation of cross-cultural data—to review my raw footage. Collier grew up in the Southwest and spent part of his childhood in Taos Pueblo. An accident in his youth (he was run over by a car) destroyed much of the auditory part of his brain, which may have been a blessing in disguise, because it forced him to rely on visual information in a way in which most of us are incapable of perceiving. Collier has produced truly remarkable still photographs of native peoples in North and South America, and he was so talented that I thought he might be permanently wedded to the still-camera format. However, using my time-motion analyzer to review my movies, he saw precisely what I had seen and more. Impressed by what a simple, hand-held, super-8 movie camera could do, Collier soon began to record on moving film the events that he could not capture with stills. Along with his gifted son, Malcolm, he has produced some remarkable books describing the recording of what was actually going on in Native American classrooms being taught by AE whites, by Indians trained in white schools, and by Indians and Eskimos who had no formal training. These studies covered a wide range of groups from the Indians of the Southwest to Eskimos in Alaska.[5] Again, the Colliers found rhythms. A quite remarkable but not unexpected discovery was that the teacher determined the rhythm of the classroom. Classes taught by Native Americans who had not been trained by white educators had a rhythm close to that of natural, relaxed breathing and ocean breakers (i.e., about 5 to 8 seconds per cycle). That is much slower than the frantic quality of a white or black classroom in the urban settings which most American schoolchildren encounter today. Native Americans who had been through U.S. educational mills produced rhythms that were in between. The Colliers' material made me realize that it was only when the Indian children were immersed in their own familiar rhythm that they felt comfortable enough to settle down and learn.

To return to my own film footage, consider one scene from an Indian market: An AE woman from the American Midwest wearing a cotton print dress and straight-brimmed straw hat was trying to be polite and nice to people she had been brought up to look down upon. She had just approached a table full of pottery. Behind the table sat a woman from Santa Clara Pueblo. Watching the white tourist enter the scene, I had to remind myself that what she was doing might not be her fault. She looked at the Pueblo woman and smiled condescendingly. Before my eyes, on the movie screen, the microdrama began to unfold. Holding herself in, the woman began bending forward from the hips to help bridge the gap made by the table, then her arm rose and slowly straightened at shoulder height. My God! It was like a rapier! The extended finger came to rest only inches from the Indian woman's nose and then it stayed there, suspended in midair. Would it never come down? The mouth moved continuously throughout the transaction: Questions? Statements? There is no way of knowing, for this was an unobtrusive record—there were no booms or shotgun microphones, no sync sound. After a while the Indian woman's head slowly rotated away from the offending finger deep inside her personal space and an expression of unmistakable disgust covered her face. Only then did the arm come down. The tourist's body rotated and she slowly moved away, with a smug, superior look on the face. Total time thirty seconds.

Analyzing this encounter, I realized that part of the communication—the real impact of the woman's unspoken feelings—wasn't just in the pointed finger but in that extended time interval that the accusing finger was held in place—the fact that she wouldn't let go but held on almost as though she were pinning an insect to a sheet of paper.

There were more encounters, fortunately none with quite the extended intense effect of the one just described. Another tourist approached a table which was apparently unattended at the time. I watched while territorial markers emerged and were played out on the screen. The tourist got too close; it was evident that he was not well coordinated and that he might rock the card table, which was tightly packed with fragile, expensive pottery. A handsome young Pueblo matron sitting a few feet away rose from her chair, straightened her spine, slowly walked to the table and placed the extended fingertips of both hands on the table's edge. There it is: "This table is mine"—said in movement and gesture. The tourist backed away and continued his conversation. I could tell from the context that not a word had been said about what really happened in that transaction. It is doubtful that either party was aware of more than a fraction of what had transpired, or that communication was occurring on multiple levels.

The question, then, was: Could other people see these things? Could people who have not been to the Southwest or lived there for years see them too? I decided to find out by repeating an adaptation of a procedure used in various research programs in which it had been demonstrated that what people see is very much a function of what they have been trained or have learned to see in the course of growing up. Each person sees a slightly different world than everyone else, and if the people are from different cultures, the worlds can be very, very different. The question was, could students overcome their earlier conditioning and learn to see differently if subjected to a prolonged and repeated exposure of short segments of film?

In those days at Northwestern University, it was possible to hire students through one of the government's student aid programs for a percentage of what the student was actually paid. This aid made it possible for me to risk a limited amount of money on something that had not to my knowledge ever been done before. I asked the personnel office to send me the next student who came in looking for work. Of course, the people in personnel wanted to know what the student would be expected to do and what skills he or she had to have. I explained that it was more important for me to have the next student than one who had a particular skill. Actually, what I needed to know was whether students who were not trained in visual analysis and who were uninformed as to the subtleties of interethnic encounters in the Southwest could on their own, and without prompting

from anyone, see what I had seen and make the same interpretations I had made.

The first student, Sheila, was an English major. I showed her
the time-motion analyzer, demonstrated how it worked and, having assured myself that she knew how to run the machine, said,
"I want you to look at these films and keep on looking until you
begin to see things in the films that were not obvious to you at
first." Sheila, of course, wanted to know what she was supposed
to see and I told her that I had no idea what she would see, but
my only condition was that she keep looking even if she thought
she was going to go out of her mind from boredom. In the
process, I began to feel like the worst kind of tyrannical taskmaster. Two days went by and Sheila, with a worried expression on her face, stuck her head in my office. "Dr. Hall, I
don't see anything; just a bunch of white people wandering
around and talking to those Indians." I said, "Sheila, just keep
at it. You haven't been looking long enough. I know it's not easy,
but trust me." Sheila tried every dodge in the book; she even went
into my files and got out films she had not been told to review.
This was all right, because I knew that she would need a break
from time to time. Her verbal skills were no use at all; she was
learning to see things in a new way and would return to her
assigned task when she felt up to it. This process of walking
into that darkened room, turning on the projector, and going
over and over that fifty-foot film clip until she felt she couldn't
stand one more look at Indians and white people sauntering
around in the New Mexico sun lasted about three weeks. But
one day, just when I was about to despair that she would ever
see anything at all, Sheila burst into my office in an obvious state
of excitement: "Dr. Hall, please come in here and look at this
film." Clearly, she had found something. The frozen image of
the woman in the print dress was on the screen. There she was
in her cotton dress and straight-rimmed straw hat right out of
the middle of the nation's breadbasket. Starting the projector,
Sheila began to speak: "Look at that woman! She's using her
finger like a sword as though she is going to push it right
through that Indian woman's face. Just look at that finger—
the way she uses it. Did you ever see anything like it? Did you
see the way that Indian woman turned her face away as though

she had just seen something unpleasant?" Every day from then on Sheila found something she hadn't seen before in the film. At first it was difficult for her to accept the fact that what she was seeing had been there all along; that what she hadn't seen at first and what she was able to see now were the same. The film hadn't changed; she had changed.

With each succeeding student, the scenario was repeated: irritation, puzzlement, boredom, searching the files for something interesting, and then suddenly when I was about to give up, a flash: "Did you see that?" Over a two-year period all students saw the same things, and in very much the same order.

Later in New Mexico I decided to use this same procedure as a sort of interethnic test. My question was: Would Spanish Americans in New Mexico, with their polychronic time system and their related deep involvement with each other, take as long to learn to read film as the monochronic, less-involved Anglos? I was not particularly surprised when the person-oriented Spanish Americans with whom I was then working proved to be remarkably adept at reading nonverbal behavior, quickly learning that film was layered information in depth. Attuned as they were to each other's mood shifts and subtle nonverbal communication, the Hispanic subjects mastered film reading in a fraction of the time required by Anglo undergraduates. Normally a weekend was enough.[6]

During an earlier research project on the subject of nonverbal communication as a factor in interethnic encounters, we discovered that ethnic blacks are even more attuned to the significance of subtle body cues than the New Mexico Spanish. It must come as somewhat of a surprise to people raised in a word culture to discover the great differences in the ability of ethnic groups to read nonverbal cues. How unfortunate that these skills are never tested in standard intelligence tests.

In a culture such as our own, with a time system like ours, people are conditioned—with rare exceptions (teenagers who see a movie twenty times)—to viewing a single performance. Even reruns on TV are avoided and only viewed if there is nothing better to do or if the movie is a classic revival. We demand variety and shun what we have already seen. This introduces a certain superficiality, a certain lack of depth that leads to dissatisfaction with the simple things of life. It was this pattern that had to be overcome in Sheila and my other students. Repetition is something few Americans are trained to appreciate. Perhaps this is why the invisible rhythm is not widely recognized in our culture, because if there is one thing that is the essence of rhythm it is that the intervals are repeated. Our real rhythms are therefore buried and must operate out of awareness. They can only be seen on stage and screen when conveyed by talented performers, or in microanalysis using a time-motion film analyzer.

Interpersonal Synchrony

It is hard to write about rhythms in English. We don't have the vocabulary, and the concepts aren't in the culture. We in the West have this notion that each of us is all by himself in this world—that behavior is something that originates inside the skin, isolated from the outside world and from other human beings. Nothing could be further from the truth.

Have you ever had the feeling that, under certain circumstances, particular people have a good or bad effect on what you are doing? The chances are that you are right, and that you should pay attention to this feeling. Discussing this rhythmic web with a friend, I found myself listening attentively as he described an example from his own experience: "Our family was at breakfast, and my daughter, who is very bright and unusually sensitive to mood and the microcosm of human transactions, was sitting diagonally across the table from me. Back against an adobe wall, I reached out to pour myself a cup of coffee. My fingers, without warning, simply let go of the half-full cup. Before I even had a chance to become annoyed with myself for being so clumsy, my daughter said: 'Did I do that?' Somehow, without realizing how, she had managed to disrupt a rhythm. How she knew she had done this I do not know except that we are unusually attuned to each other." William Condon, whose work will be discussed shortly, might provide some clues. It had to do with the delicate web of body rhythm that ties us together—a break of some sort, the short-circuiting of an action chain at a crucial point. The implications are almost beyond belief—both for good and for evil.

Condon says that when people are talking, the two central nervous systems drive each other. Of course, there are certain people who have a talent for breaking or interrupting other people's rhythms. In most cases they don't even know it, and how could they? After all, it's other people who are having the accidents, breaking and dropping things, stumbling and falling. Fortunately, there is the other kind of person: the one who is always in sync, who is such a joy, who seems to sense what move you will make next. Anything you do with him or her is like a dance; even making the bed can be fun. I know of no way to teach people how to sync with each other, but I do know that whether they do or not can make a world of difference in a relationship.

While personality is undoubtedly a factor in interpersonal synchrony, culture is also a powerful determinant. Polychronic people must stay in sync because if they don't the kind of dissonance alluded to earlier is bound to occur. I discovered this with my Spanish friends and neighbors in Santa Fe a number of years ago. While we were building a house and were all working in close proximity, it became clear how much faster and more adept the Spanish were. It was as though our small work crew was a single organism with multiple arms and legs that never got in each other's way. Synchrony of this sort can make a difference in life or death situations or in whether or not people get hurt on the job. If two or more men are lifting a heavy roof beam while standing on a wall, they must move as a unit. If they don't, one of them ends up supporting the whole load and is pushed off the wall. What I am describing is a simplified, slower-moving version of what one sees on the basketball court during championship games or when a good American jazz combo really begins to "groove," with the players constituting a single, living, breathing body.

One can observe coordination of this sort in Japan, where people work in close proximity to each other and live and breathe as a group. Even vice-presidents of large firms such as Toyota[7] frequently share offices to facilitate decision-making via consensus and remain clued into each other at all times. The end result has made a major contribution to Japanese dominance in the world's industrial and product line markets. In the AE pattern, the office is part of the symbol system in the prestige and ranking hierarchy. American executives seal themselves off from each other—to compete better. Corporate vice-presidents in the United States have to make a real effort to get together because the American system is one in which the status of the individual is closely tied to the space which he occupies. It is no accident that we refer to such things as a "badge of office"!

Status is important. However, in Japan, the markers are different. The group is more important than the individual; Japanese groups live and work and play as a unit. Toyota's assembly line teams start the day doing exercises together, then they work together, take their breaks together, eat together, live next to each other in a company compound, and even vacation together. In the past, I have watched them work in incredibly small places. I have been impressed by how they move in synchrony, a necessity in cramped quarters. I would predict that when faster methods are developed for studying synchrony, the close relationship between cultural homogeneity, polychronic decision-making, and close proximity of the members of working groups to each other will be clearly demonstrated. Actually the means are already available for studies of this sort, using the relatively simple methods described earlier. Even without these studies there is no doubt in my mind as an experienced observer of synchrony that the Japanese are more in sync on the job than Americans or Europeans. One clue is that the Japanese are more aware of synchrony than the average Westerner. Those tremendous Sumo wrestlers, for example, must synchronize their breathing before the referee will allow the match to begin, and the audience is fully aware of what is happening. In this same vein, Japanese who are conversing will frequently monitor their own breathing in order to stay in sync with their interlocutor!

Love, Identification, Synchrony, and Level of Performance

George Leonard, who has studied the rhythms of people, is convinced that nothing happens between human beings that is not reflected at some point in a rhythm hierarchy.[8] John Dewey was also interested in rhythms. In his book *Art as Experience* he states, "a common interest in rhythm is still the tie which holds science and art in kinship." Dewey believed that rhythms pervade all the arts: painting and sculpture, architecture, music, literature, and dance.

I once had an American friend and colleague—Dr. Gordon Bowles—who was both bilingual and bicultural in Japanese and American, with a slight edge in the direction of Japanese. Gordon loved Japan and the Japanese. The two of us worked together one summer in northern New York State preparing students for study and research in Japan. Every so often Gordon would disappear for a few days and when he would return, I would say, "Gordon, you've been with the Japanese again." He would reply, "Yes, I have. Some friends came through from Kyoto and I met them in Detroit. We had a wonderful time. But how did you know?" "It has something to do with the way you move—your rhythm. For a few days you move to a different beat. Then it begins to switch around to the American pattern again. It affects your entire being!"

As one might suspect, there is a relationship between rhythm and love: they are closely linked. In fact, rhythm and love may be viewed as part of the same process. People in general don't sync well with those they don't like and they do with those they love. Both love and rhythm have so many dimensions that the rhythmic relationship to love might be easily misinterpreted.

After many years in the classroom, I noticed that if I couldn't love my students, the class didn't do well, and that the rhythm of the class refused to settle down and was constantly changing. There would be good days when a rhythm seemed to be present and others when it was not, when there were several competing rhythms. A class that is going well develops its own rhythm, and it is that rhythm that pulls both the students and the professor to each meeting. What does it mean to love one's students? It sounds out of place in a university classroom, doesn't it? I am not sure it is even possible for me to unravel and identify the multiple strands that make up this particular tapestry. The classroom can be an extension of the home. It is therefore necessary for the professor to discourage any impulses on the part of the students to cast him in the parental role. Somehow the idea must be accepted that the greatest pleasure and real expression of love on the part of a teacher is to be able to watch and occasionally encourage the talent of each member of the group to grow. Also needed is the trust to permit each to do his or her own thinking. This means that we strive to bring out the best in each other and to somehow allow the rhythm of the group to establish itself and avoid at all costs the imposition of the artificial rhythm of a fixed agenda.

On the interpersonal level, observations have been made that when a mate becomes involved with someone else, there is a shift in his or her rhythm. It's as though a third person were in the house, and in a way they are, because their rhythm is there.

Individuals repeatedly demonstrate that there are very great differences in regard to their basic rhythms. There is "fast Jane" and "slow John." They should never marry or work together. These are people who temperamentally are so far above or below the average that while they can manage with some effort to synchronize with the average person, they are not able to approach the extremes of the rhythm spectrum. Most people seem to have the capacity to get up to speed, as it were, so we don't notice them, and it is not known how much this speeding up—or slowing down—process contributes to stress. These points are common experiences shared by practically everyone. Is there a person alive who has not been either held back or tailgated by others?

As any athlete knows, after strength and endurance, success in sports is largely a matter of rhythm. The super athletes are those who "have rhythm," which is why they look so beautiful and graceful when they are performing. Motorcycle riding can hardly be considered an aesthetic, to say nothing of a rhythmic, sport. Yet the all-time motocross champion, Malcolm Smith, had rhythm and won every major award in the motocross class. It didn't seem to matter whether it was desert sand, arroyos and brush, mud, rocky mountain trails, or rough desert terrain. All the other riders were yanking on their handlebars, manhandling the machines around stones, logs, shrubbery, and bad ruts. Yet a film of Smith (*On Any Sunday,* with Steve McQueen) reveals a symphony of effortless ease. He would establish his rhythm at the beginning of the race and never deviate from it. Most remarkable was that this man who was passing everyone else did not seem to be going very fast. In fact, the other contestants, when looked at individually, actually gave the appearance of going faster than Smith. It was mind-boggling to watch a man traveling at such a leisurely pace consistently pass the furious speed demons.

George Bernard Shaw, the Irish playwright who knew so much about human nature, captured this point in an essay, "Cashel Byron's Profession," the story of a public school boy with effortless rhythm who defeated his strongest schoolmates in boxing. He went on to become a champion boxer and then a respected Member of Parliament by using effortless, faultlessly timed attacks in Parliamentary debates. An even more dramatic case is reported by George Leonard,[9] who describes the extraordinary performance of a friend going for his black belt in Aikido: "So gentle and coherent were his movements that they seemed to capture time itself and to slow it to a more stately pace . . . As the exam continued, the speed and intensity of the attacks increased, and yet there was still a sense of time's moving slowly, at an unhurried dreamlike pace." This rhythmic coherent defense was maintained by the candidate while he was being simultaneously attacked by several other students. It is paradoxical that velocity, which under ordinary circumstances would be unmanageable, appears to slow down and become manageable when the right rhythm is established.

In fact, it is a fundamental truth of Zen that straining is the enemy of rhythm. Also, whatever the performance, the more perfect the rhythm, the easier it is for another person to perceive the details of what is taking place before his eyes.

Did you ever see an unusually graceful person who lacked a natural sense of self or basic confidence? The key is in the rhythm. For those interested in confidence building, or improving performance and grace, it should come as no surprise to learn that one of the most effective and rewarding combinations is gymnastics and speech training. Gymnastics—done under the watchful eye of a true professional—is the most important element and should be given the first priority. For those who are either less confident or less energetic (apparently age is not a barrier), dancing, choral singing, playing musical instruments, even marching, contribute to body synchrony, confidence, and a general sense of well-being.

The above foreshadows possibilities for progress when enough is known about synchrony so that it can be used both diagnostically and therapeutically to improve a wide variety of disorders. My intuition tells me that depression, which is so common in today's world, may have its roots in the person who is out-of-sync in deep and basic ways. Certainly, compatibility has much to do with the degree to which individuals sync with each other.

Synchrony and Group Cohesion

It should be clear by now that it is impossible to synchronize two events unless a rhythm is present. Rhythm is basic to synchrony. This principle is illustrated by a film of children on a playground. Who would think that widely scattered groups of children in a school playground could be in sync? Yet this is precisely the case (reported here in slightly revised form from *Beyond Culture*). One of my students selected as a project an exercise in what can be learned from film. Hiding in an abandoned automobile, which he used as blind, he filmed children playing in an adjacent school yard during recess. As he viewed the film, his first impression was the obvious one: a film of children playing in different parts of the school playground. Then watching the film several times at different speeds—a practice I urge all my students to use—he began to notice one very active little girl who seemed to stand out from the rest. She was all over the place. Concentrating on that girl, my student noticed that whenever she was near a cluster of children the members of that group were in sync not only with each other but with her. Many viewings later, he realized that this girl, with her skipping and dancing and twirling, was actually orchestrating movements of the entire playground! There was something about the pattern of movement which translated into a beat—like a silent movie of people dancing. Furthermore, the beat of this playground was familiar! There was a rhythm he had encountered before. He

went to a friend who was a rock music aficionado, and the two
of them began to search for the beat. It wasn't long until the
friend reached out to a nearby shelf, took down a cassette and
slipped it into a tape deck. That was it! It took a while to synchronize the beginning of the film with the recording—a piece of
contemporary rock music—but once started, the entire three and
a half minutes of the film clip stayed in sync with the taped
music! Not a beat or a frame of the film was out of sync.

How does one explain something like this? It does not fit
most people's notions of either playground activity or where
music comes from. Discussing composers and where they get
their music with a fellow faculty member at Northwestern University, I was not surprised to learn that for him, and for many
other musicians, music represents a sort of rhythmic consensus,
a consensus of the core culture. It was clear that the children
weren't playing and moving in tune to a particular piece of
music. They were moving to a basic beat which they shared at
the time. They also shared it with the composer, who must have
plucked it out of the sea of rhythm in which he too was immersed. He couldn't have composed that piece if he hadn't been
in tune with the core culture.

Things like this are puzzling and difficult to explain because
so little is known technically about human synchrony. However,
I have noted similar synchrony in my own films of people in
public who had no relationship with each other. Yet, they were
syncing in subtle ways. The extraordinary thing is that my student was able to identify that beat. When he showed his film to
our seminar, however, even though his explanation of what he
had done was perfectly lucid, the members of the seminar had
difficulty understanding what had actually happened. One school
superintendent spoke of the children as "dancing to the music";
another wanted to know if the children were "humming the
tune." They were voicing the commonly held belief that music
is something that is "made up" by a composer, who then passes
on his "creation" to others, who, in turn, diffuse it to the larger
society. The children were moving together, but as with the
symphony orchestra, some participants' parts were at times silent.
Eventually all participated and all stayed in sync, but the music

was *in them*. They brought it with them to the playground as a part of shared culture. They had been doing that sort of thing all their lives, beginning with the time they synchronized their movements to their mother's voice even before they were born.[10] This brings us to the real pioneer in this fascinating field of rhythm.

Before the Renaissance, God was conceived of as sound or vibration.[11] This is understandable because the rhythm of a people may yet prove to be the most binding of all the forces that hold human beings together. As a matter of fact, I have come to the conclusion that the human species lives in a sea of rhythm, ineffable to some, but quite tangible to others. This explains why some composers really do seem to be able to tap into that sea and express for the people the rhythms that are felt but not yet expressed as music. Poets do this too, though at a different level.

Tedlock[12] reports something very similar for the Indians of Zuñi Pueblo. Zuñi songs are composed for each year's ceremonies. A single composer will bring a song to the kiva before a dance. He will talk about the song, sing the introductory part, and then recite some of the body of the song (the "talking about" part). If the song has possibilities, his clan brothers will go to work, editing it, cutting words, changing some and, most important, matching the lyrics with the melody. It all has to fit: the words, the melody, and the message of the song. Everything has to be right. Of the 116 songs which she recorded, Tedlock reports that less than 4 percent were considered *co'ya* or beautiful, while 26 percent were *k'oksi* or good. When songs are really beautiful or good and the audience likes them, they will ask the dancers to do them again. Like good jazz—which also springs from the hearts of people—Zuñi music is judged according to how closely it approximates the living reality of the different currents in the sea of rhythm in which people are immersed. The songs perform multiple functions: religious and ceremonial, social feedback, and social control, because they frequently describe in recognizable, unmistakable detail the actions of members of the community. In Western thought, religion is one thing and social control is another. Not so for the Zuñi (or any other Native American group I know). Theirs is a comprehensive philosophy. Religion encompasses everything and is neither set aside from life nor compartmentalized. The songs, therefore, perform an emergent, formulative function because they come from that unconscious, previously unverbalized layer representing group sentiments and beliefs. That is why the very good songs are *co'ya* (*co'ya* is congruence on all levels).

One of the differences between white Americans and Native Americans, as well as blacks, is that the latter two are closer to their music. Most blacks know where their music comes from— it comes from them. The Pueblo peoples, as well as the Navajo and other American Indian groups, recognize that a people's music is inseparable from their lives and that the songs represent an important part of their identity. This is one of the reasons they don't want strangers recording sacred songs during the public part of ceremonies. Another reason is that during a dance the audience has a function to perform. That function is to be there with good thoughts and prayers! The audience's role is to add to the dance, not to take things away.

Not only do Native Americans have a beat and rhythm all their own which is reflected in their music, but each region and town in the United States has its own rhythm as well as music. An excellent example was recently provided in the opening scene of the movie *Nine to Five*, starring Lily Tomlin, Jane Fonda, and Dolly Parton. The talented Miss Parton sings the music with ground-level shots of people's legs and feet as they walk down the street. One fantastic shot zeros in on feet and ankles, in beat, cutting to a shot of three metronomes—in sync with each other and with the beat of the city. It's only a short shot, but it sent shivers up my spine. The late Goddard Lieberson experienced the power of what I am expressing so strongly that he was motivated to spend the last two years of his life producing a two-hour CBS special, "They Said It with Music," with Jason Robards and Bernadette Peters. This was the history of our country in music, beginning with "Yankee Doodle" and the Revolution and ending with World War I and "Over There." According to Lieberson, no one had ever done this before, and I can't imagine why not. Perhaps it's because we no longer think of God as sound or vibration.

Differences in Frequency, Adumbration,
and Feedback Rhythm

Feedback is a term derived from the field of cybernetics, a technical word coined by Norbert Wiener.[13] Cybernetics is the study of controls. If we consider the problems of steering a ship, both the pilot and the automatic pilot work on the principle of correcting the natural tendency of the ship to drift away from a course established for a given voyage. Various devices and aids, such as compasses, star charts, and inertial systems, are used by the pilot to stay on course; the wind, the ocean swells and currents, as well as irregularities in the ship's hull, work to push the ship off course. The link between the forces pushing the ship off course and holding the ship on course is feedback—"information" concerning how far the ship has deviated and how much correction will be needed to bring it back on course. Human beings—in fact, all living organisms—depend on feedback from the environment—human and physical—to maintain the necessary stability in life. Part of the strategy in any feedback mechanism is to know the proper interval for corrective action. If the correction is too fast, the system becomes unstable; if it is too slow, the ship wanders wide of the mark, is brought back toward the course line, crosses it, and so forth. As a consequence, the distance traveled is longer than necessary and resembles a snake or a meandering stream. This critical correction interval, which I have termed the feedback rhythm, is a function of many things, but in humans this rhythm is culturally determined at the primary level. A little-known source of communicative dysfunction is failure to match feedback rhythms with corrective action.

In an earlier work, *Beyond Culture,* I described how the polychronic Spanish people of New Mexico kept very close tabs on each other's emotions so that even slight variations in mood were detected immediately and commented on: "Theresa, what's the matter? A few moments ago you were cheerful. Now you are sad. Is anything wrong?" This sort of quick reading can be good for group morale, particularly if the group is made up of people who get along well, but it can also lead to disastrous confrontations between younger males. The same sensitivity and quick reading of low-level, nonverbal cues coupled with hair-trigger machismo, alcohol, fast cars, and guns make for real trouble. While the Hispanics are generally more tuned into each other than are the Anglos, their short-cycle feedback on the interpersonal level makes for greater volatility. A concomitant pattern is their lack of interest in long-term planning, which is always difficult to achieve in polychronic cultures unless other critical elements are present. Things happen quickly and the consequences are commonly not considered.

The Japanese have built-in systems for keeping in touch on the emotional level. This is particularly important for teams that are working together on a daily basis. The basic patterns seem to apply no matter where one taps into the Japanese hierarchy. In the morning, the Japanese start off being formal, and as the day progresses, if things are going well, the language used becomes less formal. Dropping the honorifics (suffixes which mark status signaling where each person is in relation to the other) proceeds at a steady pace. This means that everyone is up to the minute on how things are going. Unlike the Spanish of New Mexico, the Japanese do not get technical about what is wrong because they depend a great deal on context, and people are supposed to know what is wrong. In this instance we have short-term feedback—a daily rhythm broken down into interaction segments—which keeps the members of the working and living group in tune with each other and which synchronizes the emotional tone of the group. I do not want to give the idea that all groups and all Japanese work in complete harmony; they don't. It's just that they have an ideal, a method, an appropriate rhythm, a strong drive that motivates them to move from one pole (formality) in their daily transactions to the opposite (informality), which is warm and comfortable.

What type of feedback rhythm do we find on the interpersonal level in the United States? Depending on one's class and ethnic background, there is considerable variation. Even in a diverse society such as our own there are norms, because without norms it is difficult to get enough synchrony for anything to work. This kind of behavior is not the technical, verbal, manifest, explicit type found in books or in directives from management, but rather in the collective unconscious of people across the nation.

In general, Anglos, when compared to Spanish Americans, have a very long-term interpersonal feedback rhythm. They take it for granted that there will be mood shifts. At the office, when something is wrong, it is attributed to trouble at home and vice versa. Anglos tend to avoid interfering or intervening in the lives of others. This is in part a function of the monochronic, compartmentalized time system and the reinforcing effect it has on our highly individualistic culture. People frequently feel that they are alone in the world, and that it is right and proper that they should be able to solve their own problems. Any failure to do so is a sign of weakness or lack of moral fiber. What happens when things do go wrong? At first nobody says anything, and if they do it is only after it is obvious to everyone that matters are completely out of hand. A young friend who recently quit her part-time job with a professional man who had a habit of exploiting his female help could only tell him that she resented what he did and how he ran his office and that she was going to quit. He, in turn, then felt he could vent some of his feelings about her performance. It is too bad they couldn't have conveyed their feelings sooner. In marriages, individuals can go for years before they say anything about things that have been bothering them. Occasionally, one runs into a marriage or an office relationship where feedback is reliable and quick, but my own experience, as well as my reading in the folk literature on the subject (advice columns, comic strips, and letters to trouble-shooters in the press), is that three to six months is the normal interval, but that it can take up to five years before grievances are aired. This is quite different from the Japanese and the Spanish American. Spanish feedback intervals are shorter than the Japanese, while Japanese intervals are much shorter than those for white Americans.

10 Entrainment

Entrainment is the term coined by William Condon for the process that occurs when two or more people become engaged in each other's rhythms, when they synchronize. Both Condon and I believe that it will ultimately be demonstrated that synchrony begins with the myelination of the auditory nerve about six months after conception. It is at this point that the infant can begin to hear in the womb. Immediately following birth, the newborn infant will move rhythmically with its mother's voice and will also synchronize with the voice of other people, speaking any language! The tendency to synchronize with surrounding voices can therefore be characterized as innate. Which rhythm one uses, however, is a function of the culture of the people who are around when these patterns are being learned. It can be said with some assurance that normal human beings are capable of learning to synchronize with any human rhythm, provided they start early enough.

Clearly, something so thoroughly learned early in life, rooted in the organism's innate behavior program and shared by all mankind, must be not only important, but also a key contributor to the survival of our species. In the future, it is entirely possible

that synchrony and entrainment[1] will be discovered to be even
more basic to human survival than sex on the individual level
and as basic to survival as sex on the group level. Without the
ability to entrain with others—which is what happens with
certain types of aphasia—life becomes almost unmanageable.
Boston pediatrician Dr. Barry Brazelton, who has spent years
studying the interaction between parents and children from the
moment of birth, describes the subtle, multilevel synchrony in
normal relations and then states that parents who batter their
children have never learned to sync with their babies. Rhythm is
so much a part of everyone's life that it occurs virtually without
notice. Somewhere in the process of synchrony there is a link
between the normal experience on the conscious level and the
so-called metaphysical. Only a short step separates the rhythmic
sea in which all people are entrained and some of the more
contemporary theories concerning precognition.

To return for a moment to the role of rhythm in our lives and
why it may be so necessary to be able to entrain with others: at
present, possibly because there are so few people working in the
field, there are no great widely accepted theories of synchrony.[2]
The familiar, middle-frequency range rhythms are those that can
be consciously attended, like those of music and dance, which
are universal. No matter where one looks on the face of this
earth, wherever there are people, they can be observed syncing
when music is played. There is a popular misconception about
music. Because there is a beat to music, the generally accepted
belief is that the rhythm originates in the music, not that *music
is a highly specialized releaser of rhythms already in the individual.* Otherwise, how does one explain the close fit between
ethnicity and music? Music can also be viewed as a rather remarkable extension of the rhythms generated in human beings.

In addition to music, human rhythm systems can be viewed as
covering a broad spectrum, ranging from the .02 second (beta I
brain waves) at the short end to hundreds or possibly thousands
of years at the long end. The ultra-long rhythms with which the
classical historian and theorist Arnold Toynbee was so preoccupied require hundreds of years before they are played out,
and we see their patterns take shape in the rise and fall of
civilizations.

Toynbee's theory is derived from thoughtful observation of successions of civilizations where highs are followed regularly by periodic lows. While proving Toynbee's theories is not yet feasible, it does appear that the rhythmic tempo of contemporary mass culture may be speeding up. At least there is popular consensus that this is the case. If there are such rhythms and if they are really speeding up, there may be less time than in the past to adjust to the changes that are already on us.

There has always been great coherence in nature and it would be valuable to know more about the rhythmic interrelationships. Human beings are just beginning to recognize that there may be an underlying unity. It is necessary for us to understand that "rhythm is nature's way," and it is up to our species to learn as much as possible about how these remarkable processes affect our lives.

Condon comes closer than most to the root of the matter: "There is a genuine coherence among the things we perceive and think about, and this coherence is not something we create, but something we discover . . . Ideas and hypotheses are derived from and clarified by arduous observation . . . By making or finding distinctions within the world, however, we do not break it into fragments which can never again be brought together . . . The temporal is basic and involves history. Processes have their histories. There are many histories, so that while history is pluralistic, it is not therefore discontinuous."[3]

Condon believes, as I do, that all nature (life) paradoxically is both discrete and continuous—simultaneously and without contradiction. I also maintain that nature is not restricted to the physical world, but includes man and man's productions. Nothing is excluded from nature, particularly the microrhythms that tie people to each other.

Condon, a philosopher interested in phenomenology, was originally trained in kinesics at the University of Pennsylvania by Professor Raymond Birdwhistell and later by an associate, the late Albert Scheflen, a psychiatrist. Condon quickly discovered two things: first, that he had tapped into an unpredictably rich field of study; second, that no one else at that time was willing to make the commitment or had the patience to really develop the field. The temporal logistics in research of

this sort are impressive. Condon spent a year and a half (four
to five hours a day) studying 4½ seconds of Professor Gregory
Bateson's films of a family eating dinner. He wore out 130
copies of this 4½-second sequence. Each copy lasted 100,000
viewings. There has to be a lot going on at a family dinner to
hold someone's attention for a year and a half. And there is a
lot going on, perhaps more than we will ever know.

Condon knew that if he were to follow such a painstaking
procedure, it would be necessary to develop special sync-sound
16mm movies in which each frame is numbered, a multilevel
notation system to enable the observer to record movement of
any body part and the accompanying language, as well as
specially designed time-motion analyzers that could be run forward or backward a single frame at a time. All this had to be
set up in such a way that behavior could be recorded as a
continuum. One of the features that differentiated Condon's
work from other behavioral research was that there was a continuous running record—synchronous record—of what was happening through time to the entire body, including the words
that were being spoken. What Condon really accomplished was
to identify the building blocks used in the organization of behavior. In this sense, his research objectives and mine have
paralleled each other over the years. Viewed in the context of
human behavior, time is organization. However, Condon's insights include much more. For example, the definition of the
self is deeply embedded in the rhythmic synchronic process.
This is because rhythm is inherent in organization, and therefore has a basic design function in the organization of the
personality. Rhythm cannot be separated from process and
structure; in fact, one can question whether there is such a thing
as an eventless rhythm. Rhythmic patterns may turn out to be
one of the most important basic personality traits that differ-
entiates one individual from the next.

All human rhythms begin in the center of the self; that is,
with self-synchrony.[4] Even brain rhythms are reliable indicators
associated with practically everything that people do: they
change in sleep, indicate the kind of sleep that one is having
and even whether one is dreaming or not.[5] It should not come
as a surprise, therefore, to learn that Condon has established

that each of the six different brain wave frequencies is linked to specific parts of the self-synchrony rhythm spectrum. Brain wave frequencies are associated with speech in the following way:

Delta	Utterances	1–3 per second
Theta	Words	4–7 per second
Alpha	Short words and phones (sounds)	8–13 per second
Beta I	Short phones	14–24 per second
Beta II	Phones	25–40 per second

The one-second-frequency delta wave is apparently a basic rhythm of human behavior. According to Condon, the spoken (master) phrases fall into this one second interval which bridges precisely three shorter phrases. These three phrases in turn encompass two to three words each (a total of nine words) which are made of twenty-five phones (sounds) (see diagram below). Each body movement is precisely synchronized with this four-level hierarchical series of rhythms.

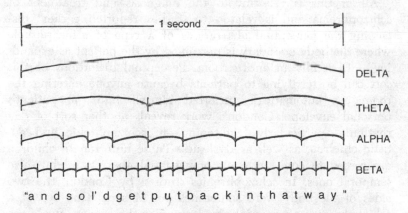

"a n d s o l' d g e t p u t b a c k i n t h a t w a y "

This fragment of speech recorded on film at 94 frames per second (almost four times normal speed) lasts for only one second. During this second the subject's arm is precisely coordinated with the theta wave pattern; her eye blinks are in sync with the beta wave pattern; the alpha rhythm is in sync with the words or vice versa.[6] Condon states: "These basic

rhythms seem to become part of the very being of the person
. . . His whole body participates in that rhythm and its hier-
archic complexities." In fact, "The *oneness and unity between
speech and body motion in normal behavior is truly awesome*"
(emphasis added). Keep in mind that these are the words of a
man who has spent the major portion of each day for the past
eighteen years immersed in the observation and recording of
these incredible patterns.

George Leonard's theory is that rhythms can account for all
sorts of psychic phenomena, and it is possible that Condon's
work may ultimately explain some of what is now seen as psychic.
This is because some form of "entrainment" is taking place whenever two central nervous systems become engaged. It is not too
farfetched, therefore, to think that some form of entrainment can
occur at a distance. Condon has demonstrated repeatedly[7] that,
when people converse, not only is there self synchrony as well
as interpersonal synchrony, but that their brain waves even lock
into a single unified sequence. When we talk to each other our
central nervous systems mesh like two gears in a transmission.

All of this is relevant to the diagnosis and treatment of
schizophrenia and is related to studies reported earlier[8] discussing the perceptual aberrations of a type of schizophrenic
where the body boundary is perceived by the patient as expanding until it fills an entire room. Perceptual distortions of this
sort can be terrifying to patients because anyone entering the
room has automatically inserted himself into the patient's
personal envelope! Condon's work reveals another sort of dis-
rhythmia—particularly frightening—in schizophrenic and au-
tistic children, as well as dyslexics. These unfortunate children
are forced to cope with not only spatial aberrations, but also
temporal ones. In schizophrenics studied by Condon, the two
sides of the body were sometimes out of sync. One eye of a
schizophrenic patient might be looking at the camera, while the
other was fixed on the therapist. Then the eyes would switch,
one eyelid closing, while the other opened.[9]

Without modern time-motion analyzers it would be impossible
to isolate, identify, and examine movements of the body occurring at fractions of seconds and below the threshold of motions
normally noticeable to the unaided eye. What was noticeable,

however, was the strain on others, when those children interacted with them. It was impossible for many of the children studied by Condon, and presumably other children like them, to entrain properly. If you can't entrain with yourself, it is impossible to entrain with others, and if you can't entrain, you can't relate.

In light of the binding nature of entrainment and given the observations in the previous chapter about the girl directing and coordinating the unconscious rhythms of the children on the playground, it is entirely possible that not only are the brain waves of people in phase when they talk, but that these rhythms are locked into the larger, more general rhythms present in whatever group we happen to be a part of at the time. And while it is difficult to prove this hypothesis at present, it could be demonstrated—or disproved—given a sufficiently gifted investigator with enough monitoring equipment to record the brain waves of individuals in groups at a distance.

A related approach, described in my *Handbook for Proxemic Research*, deals with four basic components: 1) a notation system; 2) films of people interacting; 3) a time-motion analyzer; and 4) a computer program (DATAGRF). DATAGRF was specifically designed to show when people are in and out of sync; how much they are out of sync, i.e., how much they are lagging; as well as the intensity of the transaction. This program is not perfect. It doesn't show increases and decreases of velocity, for example, but, like Condon's work, it does record events through time and is capable of handling a large number of variables in such a way that each can be compared with the others through time.

What is the effect on a group or on a single individual who is not in phase? For one thing, it can be very disruptive to the group process. I can remember being quite overwhelmed when I first made cinematographic recordings of groups of people in public. Not only were small groups in sync, but there were times when it seemed that all were part of a larger rhythm.

It has been evident for some time that the American Indian rhythm is unique. Indians talking to each other demonstrate something I have observed repeatedly but had been unable to analyze until suitable equipment became available. Theirs is a syncopated rhythm. A gesture starts with one hand, shifts at midpoint, and is completed by the other hand. The two sides of the body work together in phase with speech. AE whites not only have a left-brain culture which is linear, verbal, closed-score, and numbers-oriented, but films show that we favor one side of the body and do not carry through from one side to the other in the way that blacks and Indians do.

Pattern differences can be anxiety-provoking to people of this sort brought up to repress awareness of the body. The power of the rhythmic message within the group is as strong as anything I know. It is one of the basic components in the process of identification, a hidden force that, like gravity, holds groups together. I saw this at work in the case of one of my students —a gifted young white man who had married a black woman. He lived with her family and he had great trouble mastering even the simplest of the black walks. In fact, he never did. His wife's younger, preteen sister spent hours trying to teach him how to walk like a human being. It made her anxious and uncomfortable to watch him walk around the house like a board with hinged legs. Part of his problem was getting the right rhythm.

My own data would be termed highly suggestive. Although the differences among the three groups I studied are visible to anyone willing to look in a reasonably open-minded way, Condon's detailed data are more definitive. Describing blacks, he says: "Black behavior (some) is *beautifully syncopated* at the microlevel. For example, the black I have studied intensively will often move his right hand and arm synchronously with the initial consonant, then join in and move his left arm more rapidly with the vowel, and then both together and with reduced speed with the terminating consonant. His whole body does this . . . The white, for example, tends to keep the rhythm contained in a serial flow. There is a relative absence of a syncopated counterpoint of one side of the body in relation to the other" (italics added).[10] The reader should keep in mind that when Condon is describing what he sees, he is describing events of a fraction-of-a-second duration. Referring to the diagram discussed earlier in this chapter, this syncopation is alpha and beta brain wave level. The Native American syncopation is much easier to record, more obvious, and much slower (probably theta and delta). In fact, the tempo of normal transactions among the Indians in the western part of the United States is characteristically slower than that of either whites or blacks. How do such differences influence the way in which behavior is read across ethnic boundaries? If a single person out of sync with a group can disrupt that group, what can be the effect of fundamental rhythmic differences of the sort just described? They would not be trivial.

Life is kept on an even keel and many confrontations are avoided in AE society because people are usually able to read adumbrative cues in each other's behavior. The adumbrative sequence enables people to tell when something is about to happen and take corrective action if necessary. And while there are many types of foreshadowing, one particular category of adumbrative cues is the time lapse between saying something and doing something about it. A very significant contributor to misunderstanding between cultures is the misreading of the time between when a threat is made and when it is acted upon.

In AE cultures, if we say that we are going to do something and fail to take action after a certain time has passed,[11] it is assumed that nothing will be done. This is even reflected in law; our statute of limitations is based on this pattern. In the United States, depending on the state in which a crime has been committed, the criminal cannot be prosecuted for most crimes unless he is brought to trial within a period ranging from three to seven years. Also, a prisoner not tried following arrest within a specific period (usually sixty to ninety days after arrest) must be released. This, of course, places a tremendous burden on our courts, but it is at least consistent with the way in which our culture works, and this is more than can be said for some other areas of the law.

What is the Pueblo Indian pattern? A few years prior to the publication of this book, the Indians of one of the New Mexico pueblos caused a conflict, most elements of which were never understood by the whites. The Indians of Santo Domingo Pueblo—situated midway between Santa Fe and Albuquerque, New Mexico—told the state that they would close a state road which crossed Pueblo land unless the state made acceptable arrangements with them for the use of the right of way. Several years went by and nothing happened. Nobody bothered to talk to the Indians about the road—which served as access to a large government dam, to a real estate development on the Rio Grande, to another Indian pueblo, and to two Spanish American villages. The state highway department thought that the matter had been forgotten. Eventually, the road needed repairing, so they repaved the road without bothering to consult the Indians. Then one bright, sunshiny morning, people driving to work at Cochiti Dam found their way blocked by a steel guard rail which had been erected across the road. To be doubly sure that the barrier would be effective, the Indians had dug a tremendous trench across the old road just behind the guard rail. A sign explained that the road was on Indian lands and that the Santo Domingos were exercising their rights to close that road and that there would be NO TRESPASSING!

The whites reacted as though the Santo Domingo Indians had taken leave of their senses; the Indians couldn't understand why. The governor of the pueblo remarked, "I don't know why they were surprised. After all, those signs saying we were closing the road were stacked up against my house for a year and everybody saw them. What did they think those signs meant?" This is an excellent example of how culture teaches us to pay attention to some things while disregarding others. The signs were tangible and very real for the Indians, but the invisible time period carried more weight with the whites. Eventually, another road was built and a satisfactory right of way worked out with the Indians. In the meantime, all the people who normally used the road were forced to take a rough unimproved detour which added untold miles to their journey.

Another instance was described by Doug Boyd in *Rolling Thunder*, the story of a Shoshone medicine man in Idaho. The government was bulldozing sacred piñon trees to open up more grassland for livestock. The Indians were incensed but didn't know quite what to do, and they couldn't decide what action to take. They could, however, at least confirm what was taking place. Climbing into automobiles, the Shoshone drove out to watch bulldozers with chains between them uprooting trees by the thousands. Imagine how AE peoples would have responded.

They would have thrown themselves in front of bulldozers, screamed and yelled, circulated petitions, and filed injunctions. And there would have been no doubt in the government's mind that once the action chain had been started it would not be broken until resolved in court. Instead of becoming hysterical, the Indians just stood and looked and asked questions of a white friend who had come along to take some photographs. Years later, the Indians decided it was time to act, and when they struck they stopped the whole program in its tracks.

The fact that the Shoshone did nothing at first except talk, led the whites to assume they would continue to do nothing. Whites, on the other hand, must appear to be in the hair-trigger category when viewed by the Indians. Years of observation of patterns of this sort convince me that few people can function unless it is within the rather narrow limits of their own rhythm system. If two different systems are not calibrated, and unless a deliberate and successful effort is made to bring them into phase, the results can be disastrous. It isn't just the adumbrative segment of the rhythm spectrum that can lead to dissonance between ethnic groups; other behavioral features require our attention as well.

Interactional synchrony or entrainment across cultural boundaries is a matter not only of being in sync, but also of speeding up and slowing down at the same time that speakers move through the consonant-vowel-consonant (C-V-C) sequence. It is, as far as can be determined, a universal phenomenon that the body holds on a consonant and speeds up for a vowel.[12] When whites interact with blacks in this country, whites tend to lag on vowels—blacks do not—which makes them appear much more intense to whites. This intensity can apparently be quite threatening if one is not used to it. It may be useful to some whites to at least know these things. No analysis has been made of the C-V-C sequence used by the Pueblo Indians. I would suspect that they lag in the vowels even more than whites.

An important reason that the Pueblo Indian and the dominant AE culture are frequently out of phase with each other grows out of the differences between each culture's preconditions for an event—what must happen before a particular action can take place. For example, we assume that love will be a precondition for marriage, whereas in the Middle East, wherever marriages are arranged, other considerations take priority. If everything works the way it should, love is said to follow almost automatically.[13]

In the AE world, building a house is usually preceded by getting the land and the money in hand. The Pueblos of New Mexico have the land and, if not the money, the cooperation of one's relatives needed to build a house. However, the Pueblos have a precondition that is completely foreign to most AE peoples, one that applies to all important matters. Before a shovel of earth can be turned, all the right thoughts must be present. The Pueblos believe that thoughts have a life of their own and that these live thoughts are an integral part of any man-made structure and will remain with that structure forever. Thoughts are as essential an ingredient as mortar and bricks. Something done without the right thoughts is worse than nothing.

Think about this for a moment. What could it mean to a culture like our own? We could no longer schedule everything in advance because no one would be able to tell how long it would take to have the "right thoughts." Getting the right thoughts in one's own head as well as in the heads of others may take a long time—which can result in the overall rhythm of the culture being much slower than that of a culture that is running according to timetables set by others. Schedules, as we have noted before, set people apart and seal them off from each other. Having the right thoughts brings people together and can add to group cohesiveness and solidarity. When a Pueblo Indian builds a house, it reaffirms the group. When a white man builds a house, the last consideration in the owner's mind is reaffirming the group. In fact, building a house may even contribute to feelings of envy on the part of associates, friends, and neighbors.

The character of the rhythm is also different. Whites begin everything with capital letters or their equivalent (weddings, ceremonies, inaugurations, etc.) so that everyone will know when an event began—at what particular point in time. If one looks at these larger rhythms as though they were music, our music would come crashing forth like the starter's pistol at the beginning of a race. In fact, a reason for the strong impact of the opening passages of Beethoven's Fifth Symphony (Ta Ta

Ta Taa) is that it is so congruent with a dominant theme at the core of the AE tradition. But, think what a difference it makes when a Ta Ta Ta Taa culture interacts with one that makes very little sound at the beginning but instead "slides" into events and only moves when the thoughts are right.

There are also times when a given culture develops rhythms that go beyond a single generation, so that no one living person hears the whole symphony. This is true of the Maori of New Zealand, according to a friend, Karaa Pukatapu—a Maori who, when this was written, was Under Secretary for Ethnic Affairs in New Zealand. He described at some length how the cultivation of human talents was a process that required anywhere from generations to centuries to be completed. He commented: "What we know takes centuries, you try to do overnight!" The consequences of trying to compress long rhythms into short time periods result in AE peoples feeling that they have failed, as they do when their children don't turn out just the way they wanted. The Maori realize that it can take generations to produce a really balanced personality.

This doesn't mean that the Maori are psychologically different from Europeans when they are born, nor does it suggest that there is any innate rhythm of the type we have been discussing, only that the capacity—in fact, the drive—to stay in sync is innate and that whatever rhythms are developed by a culture will be adhered to by most members. One should never lose sight of the fact, however, that cultures over the centuries evolve their own rhythms, and since many are learned early in life, rhythms are frequently unconsciously treated as though they are innate. It takes someone with great talent and insight, like Lawrence Wylie, Harvard's Douglas Dillon Professor of French Civilization, to realize as well as actually act on the powerful effect of getting the right rhythm when learning a language (in this case French).

There is another form of entrainment experienced by many individuals for which our culture has no scientifically acceptable explanation. I am referring to Carl Jung's concept of "synchronicity," described in a paper by that title as well as in his later works.[14] In AE cultures, we are used to thinking in cause-and-effect terms. One of the ways in which this comes about has

to do with the manner in which the language as well as the time
system affects our thinking, leaving us no option but to frame our
thoughts in a linear, one-thing-after-another manner. "Post hoc
ergo propter hoc" (after the fact, therefore because of the fact)
is an old cliché in the academic world of descriptive linguistics,
used when discussing the effect of language on thought. It is
part of our AE tradition and is not the sort of thing that one
would expect to hear coming from the mouth of an old, long-
haired Navajo. Synchronicity is just the opposite. Events are
experienced simultaneously by different people in different
places so that people separated by space have been known to
experience identical sensations and emotions.

Jung said: "We cannot visualize another world ruled by quite
other laws, the reason being that we live in a specific world
which has helped to shape our minds and establish our basic
psychic conditions . . . Our concepts of space and time have
only approximate validity, and there is therefore a wide field
for minor and major deviations."[15] What was there in Jung's
experience that led him to these conclusions? There can be no
doubt that he had some remarkable experiences, some of which
are summarized below.

Jung was returning home on the train after a visit to another
part of Switzerland. He could not keep his mind on his work.
He was depressed and deeply preoccupied with thoughts of a
patient he had treated in previous years. The patient had been
dominated (submerged as it were) by a wife he should have
left but didn't. Jung checked his watch and later learned that
at the very moment that he was having these thoughts and
emotions, his patient was committing suicide. In another instance, a friend sent Jung a book on the very subject that he
was working on, which he badly needed in order to solve a
difficult problem. This case is interesting because of the added
effect on Jung, which was to help dispel a nagging notion that
because of the pioneering nature of his work he was alone in
the world. There was some validity to his feeling of being alone,
because Jung had broken with Freud (with whom he had been
on close terms) and after the break was left to continue on his
own. In a culture like ours, unless one penetrates the veil that
screens the subconscious from the conscious world, a person of

Jung's caliber is bound to feel alone. Jung's thinking, with much of which I happen to agree, is that Europeans created a conscious world in which most people live their entire lives, with little or no realization that there is another world, the world of the unconscious, which is much more in touch with culture's unifying rhythms. Jung called this world the "collective unconscious," and he seemed to draw strength from and be reassured by evidence that human beings were "in sync" on another plane.

This writer has had innumerable—many times beyond chance —experiences paralleling Jung's. Particularly prominent in my lifetime have been synchronicity in work and interests. The day before this was written a colleague from whom I had not heard for years telephoned long-distance to relay to me a matter of utmost relevance to this book. He was on the phone for an hour! In other instances, I have experienced in my own body sensations that were present in someone else's body. My only explanation is that there is a form of synchrony at the unconscious level that transcends space and sometimes time. We are used to the fact that we sync and dance quite successfully with others —in fact, sometimes we are so much in sync with them that we know beforehand what they will do next. The synchronicity of Carl Jung is only one more step down that same road. My own assumption is that as more is learned about the human rhythm spectrum and how our rhythms relate to the earth's energy fields, the explanation will become quite clear. Learning to be able to utilize syncing on this level will simply be a matter of tuning in on one's self as the Japanese do when practicing the art of Zen.

The subject of rhythm can be approached from many points of view, and it has been studied from many different angles. For another approach the reader is referred to the work of Eliot Chapple,[16] an anthropologist who has written extensively on biological clocks and their associated rhythms.

11 God Is in the Details

Arthur Miller's block-buster play, *Death of a Salesman*, made grown men weep. The play was built around the gulf separating technical, dollar-oriented business values and those of the informal (primary level–core) culture on which people construct their lives. Avarice, greed, and the failure of managers to acknowledge that there are obligations to employees (rooted in the informal system) have long been a source of tension between management and the rest of society. Miller's play describes but one example. The conflicts on which the play is based, and which give it life, lie at the crux of analogous issues that will ultimately decide the fate of our small planet, making the difference between survival and extinction if they are not resolved. In the past, humans had the luxury of catering to the egos of their leaders while following them into war. As a consequence, the human race has had thousands of years of disastrous struggles. Nations, cities, and even millions of people were destroyed but never the whole world! The question which now must be faced by the peoples of the world focuses on technical ego-centered decision making and whether or not it is an indulgence we can still afford. How does the fate of the world equate with Arthur Miller's play and the tension between informal and technical levels of culture? This is not an easy question to answer, because the answer is inherently difficult due to the way in which the human nervous system is organized. As human beings, our primary preoccupation in life is not so much conforming to the wishes and desires of others as it is to manage our own inputs in such a way that we stay comfortable and avoid anxiety. More about this later.

Staying comfortable is largely a matter of informal culture and whether one is immersed in a familiar surrounding or not. Informal or core culture is the foundation on which interpersonal relations rest. All of the little things that people take for granted that can make life with others either a pleasure or a burden depend on sharing informal patterns. There is an incredibly fluid, organic quality to the informal—as though the personal envelope which normally separates people could be expanded (as indeed it can), so that when things go well with others we are, for the moment, a single organism. The informal is intimately tied to the study of time as a cultural process. But, like Arthur Miller's play, it can also be a metaphor for larger issues.

In contrast to the technical, which is concentrated and which fragments, defines, and requires control, the informal is everywhere. In cultures like ours, it is at the informal level that the unspoken group wisdom resides. This is where most of the creative and innovative thrust occurs. The informal level is the seat of the collective unconscious and, as a consequence, the ultimate threat to the demagogue. If there is one thing that business has been slow to learn and which government has yet to learn, it is that the power, the strength, and the survival value for a people are rooted in a healthy and active system of informal culture.

What sets the informal apart is that, unlike any other form of communication, there are *no senders* and *no receivers* and *no readily identifiable messages*. Everything is in the process itself, which releases the appropriate responses in others. And when this happens, everyone is in perfect sync. In terms of our earlier discussions, the informal is very high context. It would be natural therefore that in something as low context as business and government there would be reluctance to accept as well as difficulty in understanding complex processes such as these.

Informal culture patterns are never imposed but evolve naturally in real-life situations and have stood the test of time. They come from the people. They are shared and experienced personally and are an imperative in the structure of group identity. In fact, they are what tie the individual to the group —the glue that holds the group together. Business and government continually brush them aside as trivial and idiosyncratic. It may be unfortunate but people do actually look to business for models of success and how to get things done, as they do to celebrities in the entertainment world for what life should be like and what their aspirations should be. Yet none of the above —in spite of all their power and wealth—comes even close to providing the sort of model needed by the citizens of the world today.

Because numbers can be taught, business schools have done their best to "rationalize" the management of people and resources. This has worked sometimes, but not always. There are those who have been highly critical of our business schools— even the best of them—because the schools spend too much time on numbers, and on theory, and too little on understanding people. A recent article on the subject states: ". . . a growing number of corporate managers look on them [MBA graduates] as arrogant amateurs, trained only in figures and lacking experience in the manufacture of goods and the handling of people."[1] These are flaws which critics see as reflected in management as a whole. There are also complaints about narrow perspectives and overspecialization. One can sympathize with the business schools because theory, numbers, and case histories are amenable to analysis and can be taught. Informal patterns are best learned from examples on the job. It is important to remember that business schools operate in the marketplace too and both the customer and business want procedures that can easily be grasped by managers. There is a closed feedback loop with several years' lag between business schools and business. Recently business has been saying that business schools don't teach anything about how to deal with people. Yet it has been my observation as a consultant to business that when dealing with business people in actual situations, there is *minimal* appreciation of the importance of the informal culture of the work force or the crucial role of worker morale in the whole business process. Morale is so intangible to some, while numbers are so real! Many managers "cop out," with the result that the most important part of the business is slighted. *It isn't just the cultures of other peoples that must be learned, but the informal culture of our own people!*

American law is particularly blind to the informal foundations of culture. When people address themselves to a topic as important and as close to their hearts as the meaning of length of service, they are not speaking about a casual matter which can be brushed aside peremptorily. There are literally hundreds of such patterns that go to make up our lives. Most people can't describe the rules, but they will let you know when they have been violated. Inevitably the tension between the dictates of informal culture and technical culture persists and represents one of the greatest challenges to modern management.

One such tension recently receiving publicity is the difference between male and female culture, which is responsible for some of the difficulty women have in advancing to the ranks of middle and upper management. This topic is explored by Margaret Hennig and Anne Jardim, who made an exhaustive study of twenty-five successful women from all parts of the country who had succeeded in upper management circles. Their interviews, reported in their best-selling book, *The Managerial Woman*, were extensive and insightful. Their results confirmed a point often vociferously denied by some of the more militant of the feminists, that men and women do have different cultures— informal cultures, that is. Men's and women's expectations, strategies, and attitudes toward work and group membership are quite different. What follows is a summary of the informal, subcultural differences of the two sexes as defined by Hennig and Jardim.

Because they have different informal cultures than men, one would expect women to approach their jobs differently with regard to time. This is precisely the case. Men evaluate the long-term career effects of everything they do; women don't. Women are much more likely to see the job as an isolated activity into which they pour all their efforts, but they are slow to see the career implications of what they do and how they treat people in the performance of their jobs. Men take the long-term view and tend to put up with difficult personalities, particularly those of their bosses. Men think: "This job is just one of a succession of jobs in a lifetime career, so why get myself in a turmoil because someone is hard on me?" Nowhere is the difference between women and men more pronounced than in how the two sexes handle the present in its relationship to the future. Women separate the job and the career; men don't. Men see the job in the present as well as in its career context. Women separate the two, and their emphasis is on getting the job done *now* without reference to the career. Men fail to distinguish between career in a job sense and personal goals; career is an integral part of a man's life. For women, a career is one thing and one's personal life something else. The woman's role in the more traditional sense is still centrally located in her life, whereas for the man it is not the fact that he is a man but the career that is central. With men it is what they do and with women it is what they are. Just look at the deeper meaning behind that question, "What do you do?" or "What does your husband/father do?" Asking this question of a woman can seem to be a non sequitur, particularly if asked by a man. All of this places a handicap on women who are working in organizations with men. It is taken for granted that men will devote their lives to their careers, but women have to prove that their careers won't be interrupted and that their commitment to the career is long-term. When faced with a new or difficult problem, men will ask, "What's in it for me?"—meaning, what are the long-term implications and what effect will this have on my career?

Men are brought up to be team players and women are not. There are manifold implications of this one great difference. Men are used to the fact that they might not like some members of the team, but they will not show their feelings because the team, as well as their future, would suffer if they did. Women are more likely to take things personally. Though not explicitly stated, the thread runs throughout Hennig and Jardim's book that individual time is one thing and team time is entirely different. The team takes precedence over everything else, which is why families have often been left out when businesses felt free to transfer personnel without reference to the welfare of the family. Only the team counted. Men who are brought up to be team players and managers must plan ahead. If they do not, the team can suffer. This does not mean that women can't plan or that men are better planners—in fact, many women are superior planners. It only means that women have to get used to thinking about planning ahead when they are in charge, and in Hennig and Jardim's terms, they must begin to ask themselves, "What's in it for me?"

There is a paradox in this question. On the surface it sounds self-serving. The paradox is that the question originates in male culture and acceptance of the validity of the question on the part of women who are managers represents an acceptance of the reality of male culture and, by implication, maleness in our culture. A step in the direction of accepting one's self as a person comes with the recognition of the validity of the core of another human being, and in the switch from trying to make others conform to one's own needs to allowing them to be themselves and having to cope with whatever adjustments this may entail. The "will to power" is, after all, nothing more nor less than an extension of the need to control one's inputs—a syndrome which can mask the reality of surrounding personalities in its own all-consuming flame.

Kierkegaard was right! To grow, and even to survive, human beings ultimately cannot avoid taking the first step, even if it is the analog of an American businesswoman facing her own anxiety—her angst—and accepting as valid the question her male colleagues have been asking for years, "What's in it for me?"

Apart from the inherent interest and timeliness, there are further implications which can be drawn from Hennig and Jardim's study. If differences as deep and as significant as these exist between the male and the female versions of our own culture, if this much is taken for granted in the behavior of people, just think of the effect of such differences on the international level!

It seems to be particularly difficult for the men and women who run our nation to grasp the fact that how culture molds behavior significantly influences what happens in the world. In the sense that it is used here, culture is almost totally divorced from the political process. There are ideologically neutral differences among the peoples of the world: there are monochronic and polychronic time systems, high and low context cultures, there are open and closed scores, long-term time and short-term planning, centralized and decentralized decision making, and individual and group performance on the job—all of which can be changed. If Margaret Mead's people of Manus[2] could sit down and deliberately redesign their culture and bring it in line with the twentieth century, we should be able to do the same.

But why bother to try to understand, to empathize, to learn somebody else's culture? Why bother to learn a new set of rules and new ways of communicating? Isn't the job too subtle, too complex, and too ill-defined? Perhaps. But the rewards can be very great, and the alternatives are unthinkable. First we must be willing to admit that the people of this planet don't just live in one world but in many worlds and some of these worlds, if not properly understood, can and do annihilate the others.

Time can be a metaphor for all of culture. And though we have said virtually nothing about physical time, there is one physicist, I. I. Rabi, who does have something to say. Addressing himself to the matter of time, the Columbia University Nobel Laureate says: "The real answer was given only in this century by Einstein, who said, in effect, that *time is simply what a clock reads*. The clock can be the rotation of the earth, an hourglass, a pulse count, the thickness of geological deposits, or the measured vibrations of a cesium atom" (italics added).[3] They all have one thing in common: each is a physical mechanism. Much of what has been discussed in this book is consistent with Einstein's and Rabi's statements. However, culture's clocks add dimensions to physical time, since each clock represents a particular type of organization. Like the elaborate astrolabes of the Renaissance, which were working models of our solar system, cultural models of time are also models of everything else in that culture. The metaphor of the astrolabe is worthy of further examination. It is as though each culture had its own model of the universe and lived in terms of that model. Furthermore, in at least some instances the models are so designed that they can literally annihilate each other if they overlap or are too close. As is the case with monochronic and polychronic time.

Support for this view comes from an unexpected source, Carlos Fuentes. Speaking to a college audience, the Mexican author and literary spokesman for developing countries in Latin America said: "The final question of time [is] whether we shall live together or die together . . . The West has been in love with its successive linear image of time . . . It has condemned the past to death as the tomb of irrationality and celebrated the future as the promise of perfectibility."[4] According to Fuentes, our denial of the past has led to the degradation of morality and the denial of the lessons of the past. Denial of the rights as well as the reality of other cultures is another of the consequences of Western time concepts. As Fuentes says, *We shall know each other or exterminate each other* (italics added).

Fuentes has clearly identified our dilemma and, as is typical of polychronic, highly situational logic, some of the links in his chain of arguments are missing. Nevertheless, Fuentes knows the two worlds of which he speaks as well as anyone on the globe; his views cannot easily be dismissed. My only quarrel with his argument concerns his view of how we Americans look at the future. The future in the United States is a dream. Some make the dream come true, others do not. My point is that the future is not actually real to us. If it were, how could we do such terrible things to others and to our environment? And how could our government and our businesses act so blindly, denying the reality of other cultures and, in so doing, alienating the world because of cultural ineptitude? To us, the future seems either extremely narrow or else very short-term.

Observing my countrymen over the years, I have noticed two things which stand out: our warped and inadequate view of the past and the future, and our failure to acknowledge the reality of internalized time—our own time. Time is all we have in this life, and it is my belief that life can be richer and more meaningful if people were to know more about time as it affects them personally. Then perhaps the future would begin to take on some reality and we might begin to act more realistically.

In this book, I have done my best to sketch the outlines of what will someday be an active, important, major field of study, with significance to everyone. Why do I believe that the science of time will assume greater stature in the future? There are many reasons, such as the fact that humans in all parts of the earth have been involved with time from the very beginning. If Marschack's theories are correct,[5] records of the seasons and the phases of the moon engraved by Acheulean hunters on the ribs of Ice Age mammoths represent mankind's first move in the direction of science—the earliest extensions of the human brain. Much later, in the Bronze Age, Stonehenge[6] was only one of hundreds if not thousands of early devices built to record and forecast the movements of the sun, moon, and planets. In those days, all people lived in time and, one assumes, were not as alienated from time as are many today.

The study of time has led the human species out into the universe, down into the heart of the atom, and is the basis of much of the theory concerning the nature of the physical world. In addition, it has held the attention of philosophers and psychologists, who have tried to define the nature of time as well as the experience of time.

In the second half of this century, the subject of biological clocks marked the first demonstration that all life is regulated internally and externally by rhythms synchronized with nature. Although there were only a few who recognized time as culture,[7] the study of time as a product as well as a molder of the human brain in the cultural sense was not reported until well into the second half of this century. While the study of micro time and primary level, out-of-awareness time came even later,[8] both William S. Condon's[9] pioneering work on synchrony as well as my own studies on time as an out-of-awareness system of communication cry out for continued research.

Condon's work in particular adumbrates a cultural stage, when it will be possible to make short film or TV sequences of people interacting in public—random samples—that will provide data on the degree of stress people are experiencing. The index of synchrony and dissynchrony will be as informative as samples of the blood. How people synchronize could also be used as an accurate index of acculturation. The Colliers' studies of classrooms of Native Americans and Eskimos also show great promise as a means of measuring the coherence and success of the learning environment.[10]

Much remains to be investigated about time as an organizing frame for life. Basic systems such as monochronic and polychronic time patterns are like oil and water and do not mix under ordinary circumstances. In a schedule-dominated monochronic culture like ours, ethnic groups which focus their energies on the primary groups and primary relationships, such as the family and human relationships, find it almost impossible to adjust to rigid schedules and tight time compartments. This country could do much worse than follow the example of former Congressman Ben Reifel, a Sioux Indian, who taught his people technically how to be on time for school and buses on the reservation.[11] Reifel realized that it is not enough to tell polychronic peoples to be on time or to plan ahead. Time in this sense is like a language and until someone has mastered the new vocabulary and the new grammar of time and can see that there really are two different systems, no amount of persuasion is going to change behavior. The writer Richard Rodriguez[12] has much to say about the importance of teaching language and culture in the schools. The point is that until now the schools lacked even a framework or theory for describing primary level systems.

Human beings are such an incredibly rich and talented species with potentials beyond anything it is possible to contemplate that from the perspective of this writer it would appear that our greatest task, our most important task, and our most strategic task is to learn as much as possible about ourselves. At present, it would seem that most of the world's capitals are ruled by Stone Age mentalities using Stone Age models of what the human race is all about. If the insights gained from the study of individuals trying to cope with life mean anything at all, it is that there is a direct relationship between the unvoiced picture that people have of themselves and their view of human nature.

My point is that as humans learn more about their incredible sensitivity, their boundless talents, and manifold diversity, they should begin to appreciate not only themselves but also others.

One hopes this will ultimately lead to lessening our tendency to subjugate or stamp out anything that is different. The human race is not nearly enough in awe of its own capabilities. My picture of the future is not so much one of developing new technologies as it is of developing new insights into human nature.

This book has taken one little corner of human nature and put it under a microscope. What I see is a whole new dimension or set of dimensions to be explored. God really is in the details. And I for one do not think for a moment that He intended us to blow each other off the face of the earth.

APPENDIX I

A Map of Time

When one looks at the Time mandala several things become apparent. First, there are four pairs in which the categories appear to be functionally interrelated: 1) sacred and profane, 2) physical and metaphysical, 3) biological and personal, and 4) sync time and micro time. Second, the time positions on the opposite side of the mandala also seem to bear a special relationship to each other. Sacred time and personal time are personal, and from what little is known of the metaphysical it would seem that rhythm is shared with sync time in both (see chapter 10). These common elements, such as rhythm, are links connecting the different kinds of time. Third, the two axes going from lower left to upper right and upper left to lower right set things apart in other ways: group, individual, cultural, and physical. Fourth, the left side is explicit and technical (low context) while the right-hand side is situational (high context). All of this suggests that there are clusters of ordered relationships between the different kinds of time.

The mandala also makes it possible to categorize different historical periods and cultures. The Hopi, for example, traditionally lived almost entirely in a world of sacred time. The four categories on the "group" side of the line are contained in and treated as a single capsule. Awareness of sync time is more developed in Black Africa than in AE cultures. One gets the impression that in the subcontinent of India the metaphysical and the sacred are fused into one. In the United States we make few distinctions between profane time and micro time. It appears also that if one culture emphasizes a particular segment while another emphasizes a different one, the results can be extraordinarily significant.

A MAP OF TIME

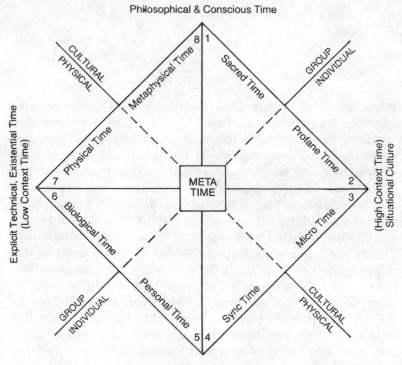

Note: To discuss complementary systems it is necessary to invoke Meta Time, which is where the integrative concepts are located.

APPENDIX II

Japanese and American Contrasts, with Special Reference to the MA

The closest one can come to understanding Japanese time is to approach via the route of MA. MA is time-space. The two cannot be considered separately. Like everything else, and particularly Zen, MA does not lend itself to technical description. MA apparently underlies almost everything and is an important component of communication. Years ago I made the observation that in the West we pay particular attention to the arrangement of objects, and in Japan it is the arrangement of the spaces—the intervals, MA—that are attended. In speech this means that it is the silences between words that also carry meaning and are significant. Americans in meetings or when giving talks and lectures are famous in Japan for failing to take account of MA when structuring their presentations. As a consequence, American speakers to Japanese audiences give the impression of tailgating the audience because the audience never has a chance to catch its breath and think about what is being said, which high context communications patterns demand. Is it any wonder that Japanese and American audiences experience the same presentation differently?

MA is much more than a silence between events (our interpretation) or events punctuated by silences. It is difficult for the Japanese to explain MA, because MA is part of the root culture of Japan (informal, out of awareness, primary culture). I have yet to read or hear an explanation of MA that would satisfy the Westerners' low context need for details. It may very well be that, as in Zen and archery or Zen and swordsmanship, one has to go through the experience to begin to understand it. My point is that our own stereotypes keep getting in the way. All peoples of the world look at that world in their own way. The world they see is one which they have created—which is why it all looks so familiar. It is only the rarest of human beings who understands the complexity and the beauty of the intercultural process. Japanese culture is contexted in so many ways and at so many different points that it is virtually impossible to describe to the Westerner. The description that follows is drawn largely from the catalog of Arata Isozaki's exhibition "MA" shown at the Cooper-Hewitt Museum in New York.[1] It is as good an example as one can find in which the nature of the contexting is made somewhat explicit. It is not important that the reader feel that he has completely understood the "meaning" of the different sides of MA, because I doubt that that is possible. What he should look for is how those concepts are stated, how they relate to each other, and how one would have to be able to think to have produced such a system.

MA is categorized as nine different varieties of experience: *Himorogi, Hashi, Yami, Suki, Utsuroi, Utsushimi, Sabi, Susabi,* and *Michiyuki*. The nine varieties of MA can be likened to nine chapters of a book or nine scenarios written by different authors. The basic themes remain fixed, but the treatment is different each time one hears it. The whole story is inherently Japanese. There is so little that is familiar to the Western mind that one is left with the conclusion that here is a system of thought that must trace its origins thousands of years in the past with little or no influence from the West.

Himorogi stands for two things: the sacred descent place of the Kami (the original pre-Buddhist, pre-Shinto Gods), and the exact moment when this occurred. *Himorogi* is reminiscent of the "big bang" theory of the creation of the universe. In Japan, the exact moment of anything is important. It is a little bit like concentrating the essence of the event into a momentary flash, which is just the opposite of the Hopi notion that many important things require repeated small ceremonies spread out over time. Our own creation myth as told in the Bible required six days, and then a seventh so that God could rest, thus establishing at the outset a distinction between working time, which was profane, and God's time, which was sacred. Symbolic representations of *Himorogi* can be found in the hundreds of temples, shrines, and gardens that dot the urban as well as the rural landscape in Japan. The form is usually four poles marking the boundaries of a square with a post in the middle. All four posts are enclosed by an encircling rope or cord. Another version is a raised platform with a tree or shrub in the middle. Two bundles of straw are suspended from the middle of the tree's trunk.

The Japanese are thus reminded daily of multiple links with the past and the importance of the exact moment. In Japan, time and creation partake of the same process. It is the presence of shrines in all districts that provides a constant visual reminder that the deep past is always present. Even in the middle of the busiest, most modern sections of Tokyo there is a hundred-meter square dedicated to the spirit of Masakado Taira, a warlord killed in a tenth-century battle. It would be foolhardy to suggest moving the monument, even though the space on which it stands is reputedly worth $10 million.

Hashi means "to bridge." It underscores the bridging function in both time and space and gives buildings a special significance which is almost sacred. *Hashi* also means the space between two things (the time between two events) and implies dividing up the world. *Hashi* also stands for edges, spaces between, and bridging. This whole book is an exercise in *Hashi*.

Yami is the world of darkness from which the Kami come and to which they return. Traditionally, the Japanese believe that the Kami permeated the cosmos and were conscious of the sun, whose movements divided time and space. The sun created day and night and life on earth. MA is maintained in absolute darkness and the word *Yami* combines the meaning darkness *(yo)* and the transition from darkness to light *(yamu)*. *Yami* therefore recalls the image of the entire universe. One sees this metaphor also in the design of the Noh stage which has a small forward

part for "this" world and a larger part for the "other" world—
the world of the spirits, and a *Hashi* bridging the two worlds.
The world of the present and the world of the dead are much
closer in Japan that they are in the West.

Suki MA means aperture, but because of the etymology of the
word it carries a connotation of "like" and the French concept
of "chic" *(furyu)*. *Suki* is difficult to relate to Western concepts
and while I am not certain about this, it is possible that our
metaphor "a window of time," a "window of opportunity," or
Alexander Haig's "window of vulnerability"[2] have some relationship to the concept *Suki*. Writing about *Suki*, the authors
of the catalog of the MA exhibit discuss at some length the
origin and evolution of the teahouse for the tea ceremony.
Modeled originally after a simple workman's hut, the teahouse
symbolized a reaction against the pretentious architectural style
of the sixteenth century. The tea master's arrangement of the dis-
play of the instruments used in the ceremony on the *tana*
(shelf) is another reminder not only of the past and tradition
but also of the art of relating and communicating to other human
beings in subtle, symbolic ways. Some tea masters were highly
regarded because they made each arrangement according to the
character and preferences of the guest.

The *tana* was situated in the *yoko* (alcove), from which the
idea of the *tokonama* originated. The *tokonama* is the most important place in the house and is used to exhibit particularly
beautiful scrolls or art objects, as well as selected objects appropriate for the season or a special occasion. The guest is seated
with his back to the *tokonama*. The *tokonama* in its expression
of simplicity can symbolize the giving up of other things, such
as pretentious ways; or it can symbolize subtlety, art, the honor-
ing of guests, and the seasonal reminders of the passage of time.
Again we find the past and the present united in a single symbol.

Utsuroi encompasses the whole process of change. Typically,
this starts with the beginning: *Utsu* (vacuum) *hi* (activity of
the soul) making *Utsuroi* (change). Two important themes are
present in *Utsuroi*, which is related conceptually to *Himorogi*—
the exact moment when the Kami inhabited the void as well as
changes in nature over time. Time and nature are closely associated in Japan. The Japanese visual metaphors, as one would

expect, differ in many ways in the context of *Utsuroi*. Mind, trees, and grasses are symbolic of growth and change. Wilting leaves and flowers are also deeply significant. In the West the season predominates in our imagery of time. In Japan it is what happens during these seasons that symbolizes time. Again, there is a shift from the larger picture to the specifics that reinforce patterns mentioned earlier. The effect of the ties to nature cannot be overstated. Everything in nature reminds people of time. Waves and currents in the ocean with their constant motion are symbolic of eternity.

Utsushimi stands for the physical projected into reality and *Utsushimi* MA is the place where life is lived—the house or home. The home is a small model of the universe incorporating altars and places for Buddha and other Gods. *Utsushimi* relates to *Utsuroi* and *Suki* and also note the absence of clear-cut categories of the type found in the West.

Sabi—like *Utsuroi* and *Himorogi*—evokes images of "the precise moment," but it includes something else, another inevitable force in life: the process of death, decay, and the life cycle. *Sabi* is the thought that everything passes through stages, from birth to decay. *Sabi* invokes feelings of dissolution and approaching extinction. One is reminded here that time is tied to the processes of nature and that the phase referred to is the final phase in a long process.

Susabi MA referred originally to the playing of games by the Kami. There is something whimsical in *Susabi*, something more than paradoxical. One gets an impression of the craziness of it all—violating all the rules of aesthetics—big buildings on spaces that are too small; kitsch; calligraphy plastered all over everything; and a general lack of congruence. *Susabi* may be the symbol of modern times. If so, we could use a similar metaphor in the West.

Michiyuki MA deals with pauses—pauses and stops on journeys. For example, the road from Kyoto to Toyko had fifty-three way stations. Each station was chosen because it showed the traveler a particularly beautiful view or noteworthy feature of the landscape. Similarly, the traditional Japanese garden frequently has stepping-stones which are so arranged that one has to stop and look down and then look up again and in doing so one sees a different perspective for each step. *Michiyuki* has elements of the schedule in it, except that the intervals are never the same. They are, however, preprogrammed.

Reading the above, I am reminded that there is no way to adequately translate from one world view to the other without a deeper knowledge of human nature. I sincerely believe that there are bridges *(hashi)* but that the foundations of these bridges lie deep within us at a level that the average person has yet to explore. This generalization applies to both the East and the West, though the East is more accustomed to thinking about this sort of thing.

If a Westerner could understand MA, it would give him some feeling for the inner experience of time in Japan. One should keep in mind several reservations. The Japanese live in two worlds, and since the worlds exist side by side, like electrons changing orbits around the nucleus of an atom, they shift from one to the other literally from moment to moment. The two worlds are, of course, the traditional and the modern. The modern is suffused with much of the West, but this can be deceptive because one can never tell exactly where one is.

This matter of shifting from one world to the next is paralleled in the daily relations between people who move between the world of the formal public self *(tatemae)* and the private self *(honne)*. One world is formally ritualistic and preoccupied with status, the other is informal, warm, close, friendly, and egalitarian. A measure of how one progresses in a relationship is how the transitions between the public and private definition of the situations are handled. A study of the informal timing rules of these transformations should prove highly enlightening. There is undoubtedly a timetable or, rather, a number of timetables, each of which has its meaning.

That Japanese culture consists of many transformations should be fairly obvious by now. Learning to live with things is an important and respected attribute of the Japanese. In the West, there are significant parts of life that until now have been highly resistant to change. And even though the nature of sexual relationships and marriage and living arrangements between the sexes have changed greatly in recent years, there are certain aspects of our culture that do not seem to be changing, such as our highly centralized way of making decisions, our use of negotiation as a way of accomplishing changed relationships, and the basic structure of our time system.

A recent characteristic of commercial, social, and academic life that does not augur well for stability in our culture is the extremely "trendy" nature of our society. The fact that something is new means more to us than anything. Apart from antiques, which are in a special category, that something or someone is old evokes images of the scrap heap regardless of how much inherent value remains. All of this is consistent with the extremely short-range schedules in our lives today, in business as well as in planning. The Japanese are also bitten by this bug: "We must be up to date and must have the newest."

If there is a single important deep difference between the way the Japanese and the AE cultures treat time, it is that time is imposed from the outside by Westerners. And even though schedules and values like promptness are internalized, our basic system has its origins outside the individual. In Japan, the reverse is true. Time starts inside the individual. While tight scheduling in Japan is virtually the first impression foreign business visitors receive, I can't help feeling that what we see is an artifact of our own Western civilization—a technical caricature borrowed to accommodate Europeans and Americans. Because everything changes with each new situation, the highly situational side of Japanese culture presents many puzzles to the foreigner. Everything is different in a new situation. If there is one piece of advice I would give Europeans visiting Japan, it would be to master a few basic situations and stick to those, and when a new one presents itself, get some help from a skilled intermediary. This is difficult for Americans and North Europeans, because none of us seems to accept the need to be coached.

NOTES

Introduction

1. Leonard Doob, "Time: Cultural and Social Aspects," 1978.
2. E. R. Leach, *Rethinking Anthropology*, 1961.
3. N. E. Howard, *Territory and Bird Life*, 1920.
4. Robert Ornstein, *The Psychology of Consciousness*, 1975.
5. Tadanobu Tsunoda, *Nihon-jin No No— The Japanese Brain*, 1978.

Chapter 1

1. E. E. Evans-Pritchard, *The Nuer*, 1940.
2. Frank A. Brown, "Living Clocks," 1959; W. S. Condon, "Neonatal Entrainment and Enculturation," 1979; Albert Einstein, *Relativity: The Special and General Theory*, 1920.
3. Ruth Benedict, *Zuñi Mythology*, 1969.
4. J. B. Priestley, *Man and Time*, 1964; Leo W. Simmons, *Sun Chief*, 1942; D. R. Sol, "Timers in Developing Systems," 1979.
5. Mircea Eliade, *The Sacred and the Profane*, 1959; Julius Fraser, ed., *The Voices of Time*, 1965; Edward T. Hall, *Beyond Culture*, 1977; Paul T. Libassi, "Biorhythms: The Big Beat," 1974.

Chapter 2

1. Shoseki was an early Zen master known to Zen scholars, and is mentioned in Karlfried Graf von Dürckheim's *Hara: The Vital Center of Man*, 1962.
2. In the 1930s, women's rights had not been raised as an issue, either in the United States generally or with the Navajo. Only "heads of households," as defined by the Federal Government, were employed.
3. Benjamin Lee Whorf, *Language, Thought, and Reality*, 1956.
4. Harold S. Colton, *Hopi Kachina Dolls*, 1949; Edward H. Kennard, *Hopi Kachinas*, 1938. The Hopi Kachina dolls are better known to the average white than the dancers on which they are modeled. The dolls are small replicas of the dancers.
5. Edward H. Kennard, *Hopi Kachinas*, 1938.
6. Leo W. Simmons, ed., *Sun Chief*, 1942.
7. Fred Kabotie, *Fred Kabotie, Hopi Indian Artist*, 1977.

Chapter 3

1. *Beyond Culture* by Edward T. Hall, 1976 (pp. 17–20, 150–51, Anchor Press/Doubleday), also discusses these two time systems.
2. The exceptions are the large and important minorities who trace their origins to Spain (Spanish Americans, Cubans, Puerto Ricans, Hispanos from Mexico, as well as other parts of Latin America). The P-pattern tends to be associated with binding family ties, with large groups of relatives. One wonders if it is not an artifact of informal culture in such a situation as that of almost a hundred relatives arriving without notice or on very short notice and making demands. The Jews, the Arabs, and the Spanish share close family ties and extensive networks of friends as a cultural characteristic, and, though there are exceptions, all tend to be polychronic.
3. Chapter 4 is devoted to context.
4. Edward T. Hall, *Beyond Culture*, 1976. Chapter 15 is devoted to the important subject of identification.
5. To complicate this particular conversation, the Japanese people do not like to say no and have therefore developed an incredible assortment of excuses which have the same effect. I knew that it sounded as though I was saying no in a roundabout way just to save his face; however, this was not the case. I really don't have any control over foreign publication of my books.

Chapter 4

1. Edward T. Hall, *Beyond Culture*, 1976.
2. Some of what follows was given as a talk by the author to the BDW Deutscher Kommunikationsverband in Bonn during the spring of 1980.
3. See Paul D. MacLean, 1965, for an exciting, promising new approach to the brain which explains much of human behavior.
4. The fast and slow message continuum as described here is a somewhat revised version of that appearing in Edward T. Hall, *The Silent Language*, 1959.
5. Michael Korda, *Power*, 1975, and *Success*, 1977.
6. PBS Television, Leonard Bernstein's introductory remarks to Beethoven's 6th Symphony, Vienna, Austria, February 28, 1982.
7. Reported in the Washington *Post*, column by Judy Mann, June 10, 1981.
8. See Edward T. Hall, *The Hidden Dimension*, 1966, for more on the subtle effects of space on human behavior.
9. High-rise public housing in the United States has proved disastrous in its consequences. The Cabrini-Green housing project in Chicago and the Oliver Wendell Pruitt Igoe public housing project in St. Louis, Missouri—to name only two—were such disasters that no major high-rise, low-cost public housing has been built in the United States for over twenty years. Yet the homogeneous, disciplined Chinese in Hong Kong have used high-rise public housing with great success. In London, the history of high-rises is mixed. My friend and colleague Ernö Goldfinger, architect and designer, says that when stable, homogeneous Cockney neighborhoods are put under one roof in a high-rise apartment, things go very well. It is only when mixed neighborhoods of recent arrivals to London from elsewhere in the British Isles are located under that same roof that real trouble can brew. Educators even have a percentage point beyond which the character of a classroom changes completely. The proportion quoted is 1:3. When one third or more of any group is injected into a homogeneous situation, the situation is likely to become unstable. The matter of homogeneity and heterogeneity is a volatile and complex subject that will be explored more fully at a later date. There is nothing wrong with either, provided enough is known before hand to take advantage of the positive aspects and to avoid the negative ones.
10. In the context of AE societies, marginal families, regardless of ethnicity, seem to have in common an almost total lack of "planning."

Monochronic planning on their part is usually either nonexistent or very short-range. Most live at the very edge of survival in situations that can only be characterized as precarious. This is not only wasteful, but it means that both the time pattern and the ways of integrating with the larger society are not used. To fail to take this time pattern into account in social welfare programs is inhumane and costly, and it condemns the recipients to eternal poverty and marginality.

11. This school was run on the principle that unless you treat the entire family, it does little good to treat the child. This rule applies to virtually any polychronic family.

12. The author is indebted to Dr. Gabrielle Palmer for her account of Ecuadorian time summarized here.

13. If you can depend on someone not to steal and do not lock things up, he has internalized social controls; if, however, you must depend on the locks to keep him "honest," the controls are external.

14. CII (Colonial Iberian-Indian). Colonial Central and South America were colonized by Spain and Portugal. The combination—sixteenth-century Iberian Peninsula and indigenous Indians—has produced a cluster of national cultures that, while very different from each other, share basic communication systems: the Spanish and Portuguese languages, as well as temporal, spatial, and other PL systems. The CII designation is used as a convenience.

Chapter 5

1. E. E. Evans-Pritchard, *The Nuer*, 1940.

2. Paul Bohannan, "Concepts of Time Among the Tiv of Nigeria," 1953.

3. There are, of course, categories of activities in AE cultures which can be interrupted with greater ease and less affront than others. There is no apparent logic to these categories. If someone is simply sitting and thinking, people will interrupt without hesitation (he isn't doing anything!). If someone is reading or writing letters, being waited on in a store, or even talking to someone else, he is apt to be interrupted. To forestall interruption, a show of importance in status or activity must be established, i.e., unusual importance.

4. The Quiché material in this chapter is based almost entirely on Dr. Tedlock's work. Dr. Tedlock spent three periods of fieldwork from 1975 to 1979 in the Guatemala highland village of Momostenago. The interpretations of Quiché culture in contrast to American culture are my own, although I have discussed them with Dr. Tedlock. The

reader who wishes to know more should go directly to her book, *Time and the Highland Maya*, University of New Mexico Press, Albuquerque, N.M., 1981. It is the only source I know of impeccable scholarship that still describes what the people actually did as contrasted with what they told others, who simply sat by and either watched or else just took notes. Having learned and practiced Quiché daykeeping (divination), she put herself in the insider's position, looking outward, rather than in the more conventional position of the outsider looking in.

5. Americans, even though they keep massive records, are seldom at ease with their past. On the personal level, they either: 1) try to discard it, cf. the born-again syndrome; 2) take an infantile approach and wallow in it, blaming their parents for all the bad things in their lives without doing anything; 3) deny it; or 4) romanticize it and reify it, as in the South.

Chapter 6

1. In a sense, much of my own work over the years is closer to the Zen model than it is to the Newtonian model that lies behind much of today's social science. Nevertheless, since I was reared as a Westerner, it is inevitable that I approach virtually everything from the Western point of view.

2. Ruth Benedict, *The Chrysanthemum and the Sword*, 1946; Karlfried Graf von Dürckheim, *Hara: The Vital Center of Man*, 1962; Erich Fromm, D. T. Suzuki, and others, *Zen Buddhism and Psychoanalysis*, 1960; Frank Gibney, *Japan, the Fragile Superpower*, 1979; Eugen Herrigel, *Zen in the Art of Archery*, 1971; Hidetoshi Kato, "Mutual Images: Japan and the United States Look at Each Other," 1974; Fosco Maraini, *Japan: Patterns of Continuity*, 1979; M. Matsumoto, "Haragei" (ms.), 1981; D. T. Suzuki, *Zen and Japanese Culture*, 1959; Ezra F. Vogel, *Japan as Number 1*, 1979.

3. D. T. Suzuki, *Zen Buddhism*, 1956; Eugen Herrigel, *Zen in the Art of Archery*, 1971; and others too numerous to mention.

4. Eugen Herrigel, op. cit., and Edward T. Hall, *The Silent Language*, 1959.

5. Eugen Herrigel, op. cit.

6. Robert Ornstein, *The Psychology of Conciousness*, 1975.

7. Low context communications are those in which the text is detailed, leaving very little for the imagination. See chapter 4.

8. Leo W. Simmons, *Sun Chief*, 1942.

9. Most of the martial arts are just that: arts. The swordsman and the archer are on the same level as practitioners of Zen, as the philosopher and the priest. In all classes there are masters.

10. M. Matsumoto, "Haragei" (ms.), 1981.

11. Edward T. Hall, *The Hidden Dimension*, 1966.

12. Japan is a society of obligations, such as *on*, the obligation to the Emperor and one's teachers or lords. *On* can never be repaid. Reciprocal obligations—*gimu* and *giri*—can be repaid but are different from one another. *Gimu* has no time limit and no matter how much one repays a favor or a grant, the obligation will still remain in part. *Giri*, on the other hand, can be paid in full, and there are time limits. *Giri* is a way of living and doing one's job. See Ruth Benedict, *The Chrysanthemum and the Sword*, 1946.

Chapter 7

1. Chapters 9 and 10 are devoted to rhythm and entrainment.

2. See Edward T. Hall, *The Hidden Dimension*, 1966, for a more detailed discussion of the central point or place in French culture.

3. It is important to mention that the procedures governing fiscal reporting are not widely known outside of financial circles. Talking to my friend Lawrence Wylie, Douglas Dillon Professor of French Civilization at Harvard, I learned that in spite of his years of experience in France this pattern was unknown to him, just as it would be unknown to the average American professor if the situation were reversed. It would be logical to assume, however, that a pattern such as this, on which so much depends, would crop up elsewhere in French culture. And while there are similar occurrences in the United States, it is considered very "dirty pool," otherwise why the outrage on the part of Americans when confronted with the "turn back the clock" syndrome in France?

4. The role of single-issue politics in the United States has become so prominent in recent years that it can no longer be ignored. Single-issue politics is the ultimate in absurdity as far as the general welfare of the country is concerned. If our country were a little higher on the context scale, this sort of thing would not be an issue. The question is, will single interests ultimately so weaken the nation that it cannot survive?

5. Open and closed score is a process described earlier (Lawrence Halprin, *The R.S.V.P. Cycles*, 1970; Edward T. Hall, *Beyond Culture*, 1976) in which two different strategies produce different performances.

A "score" (taken from music) can be anything from a shopping list to the program for placing a man on the moon. A closed score strategy succeeds if the performance follows the score and attains its stated goals, e.g., landing a man on the moon. An open score fails if nothing new is added. Music can be either, as a traditional form (classical—closed; jazz—open); individual musicians can violate tradition and assume either approach. Virtually anything that people do can be characterized as one or the other.

6. See Edward T. Hall, *Beyond Culture*, 1976, chapters 6, 7, and 8, for more on strategies of contexting.

7. For a description of formal, informal, and technical modes see my book *The Silent Language*, 1959.

8. John F. Berry, articles in the Washington *Post*, May 29 and June 3, 1981.

Chapter 8

1. *Homo neanderthalensis*—the precursor of modern man—(70,000–37,000 B.C.) "used symbols, used red ochre, had ritual burials (even with flowers) . . ." (Alexander Marschack, "Ice Age Art," 1981), all strong indications of a belief in a hereafter as well as the beginnings of religion.

2. Alexander Marschack, *The Roots of Civilization*, 1972.

3. The topic of extensions was discussed at length in my book *Beyond Culture*, 1976.

4. A work in progress is devoted to this extraordinary subject.

5. Alfred Korzybski, *Science and Sanity*, 1948.

6. The earliest recognition of this process is in the Bible, when the Israelites were told not to worship idols.

7. Margaret Church, *Time and Reality: Studies in Contemporary Fiction*, 1963. (This is one example.)

8. *Washington Post Magazine*, November 9, 1980, article by Walter Shapiro.

9. Stanley L. Englebardt, "The Marvels of Microsurgery," 1980.

10. This topic is discussed under imagery in Hall, *Beyond Culture*, 1976.

11. Paul Pietsch, *Shufflebrain*, 1981; Karl H. Pribram, *Languages of the Brain*, 1971.

12. The brain is an extraordinary organ in which all of the parts interact. For a brief summary, see chapter 12 of Hall, *Beyond Culture*, 1976.

204

NOTES

13. Edward T. Hall, *The Silent Language*, 1959. This is also the sort of difference between families of cultures that lends itself to rigorous testing.

14. Maria W. Piers, Erickson Institute Outrider, #18, Fall 1980.

15. Jean Piaget, *Time Perception in the Child*, 1981.

16. Franklin P. Kilpatrick. See notes 22 and 28 below, also my books *The Hidden Dimension* and *Beyond Culture*.

17. Jean Piaget, *The Child's Conception of Time*, 1969.

18. Jean Piaget, *The Child's Conception of Space*, 1956.

19. Patricia Carrington's *Freedom in Meditation*, Anchor Books/ Doubleday, 1977, is one of the most balanced and scholarly of the numerous reports on this subject.

20. Keith Floyd, "Of Time and Mind: from Paradox to Paradigm," 1974.

21. Note that the inherent logic of the central nervous system is not the same as Aristotelian logic. Inherent logic is somewhat like the logic of topology. It is a logic of relationships in which forms may change but relationships remain constant.

22. The totality of transactional psychology is devoted to this process. See Franklin P. Kilpatrick, ed., *Explorations in Transactional Psychology*, 1961.

23. Edward T. Hall, *Beyond Culture*, 1976.

24. There are many types of schizophrenia, and this example describes part of the symptomatology of only one type. However, this type is not unique to the United States. In a series of interviews with Dr. Paul Sivadon, a Belgian psychiatrist practicing in France who specializes in a sort of total environment therapy, I learned that the symptoms described were well known to him and sufficiently common for him to develop a unique way of treating them. He simply gave the patients more room than they could possibly fill.

25. Alton De Long, "The Use of Scale Models in Spatial-Behavioral Research," 1976; "Spatial Scale and Perceived Time-Frames," n.d.; with J. F. Lubar, "Scale and Neurological Function," 1978.

26. These results are entirely consistent with Keith Floyd's (op. cit.) conclusions based on brain-wave studies of meditating patients. In fact, Floyd states: ". . . what we think of as time is merely a function of one's basal brain wave rates, a convenient and fascinating fabrication of the conscious mind."

27. Alton De Long, op. cit.: 125, 190, 38, and 96 subjects were used, i.e., 125 subjects for the $\frac{1}{24}$ environment, 190 for the $\frac{1}{12}$ environment, and so on. Each scale was represented by a number of settings: waiting rooms, living rooms, reception areas, etc.

28. Alton De Long discovered that judgment of time—as one would expect—is independent of the higher cortical functions and that if his subjects tried to judge time intellectually the whole experiment was invalidated. Which points up some of the complexities of research of this sort. Some procedures work very well with the higher centers of the brain, others, like space perception, do not. In fact, Kilpatrick et al. (1961) demonstrated that the highly integrated processes of space perception are independent of conscious cortical functions: that intellectual knowledge that a room was distorted had no effect at all on how the room was perceived. Anxiety, however, was a different story. Anxious subjects held on to their perceptual distortions longer than normal subjects. Note that all "experienced" measured times are related to 30-minute base lines. For the $\frac{1}{12}$ environment, the mean elapsed time (judged as 30 minutes) was 2.44 minutes or $\frac{1}{12.29}$. That is, they are $\frac{3\frac{4}{100}}$ ths of 1 percent out of agreement or about $\frac{1}{3}$ of a percentage point. The $\frac{1}{24}$ scale yielded an elapsed time of 1.36 minutes ($\frac{1}{22}$ instead of $\frac{1}{24}$). For the $\frac{1}{6}$ scale, the elapsed time was 5.01 minutes ($\frac{1}{5.99}$ instead of $\frac{1}{6}$) or a deviation of $\frac{1}{10}$ of 1 percent. For the statistically inclined, De Long reports significance at p .0005 level.

29. Gannett News Service reporter Dave Schultz, writing about the 1980 Winter Olympics (dateline Lake Placid, New York), describes at some length how Debbie Genovese—No. 1 woman on the U.S. luge team—goes through a process she calls "pre-visualization": "You close your eyes and think about every inch of the course and what you are going to do . . . You think about the start and each curve, where you are going to enter a curve, where you will leave the curve. You run the race in your mind all the way to the bottom of the hill. It should take you as long to think your way through the race as it will to run the race . . . It's amazingly accurate."

Chapter 9

1. Rhythm will, I believe, soon be proved to be the ultimate dynamic building block in not only personality but also communication and health.

2. Proxemics is the study of man's use of space (including his distancing and territorial habits) as a special elaboration of culture. Proxemic observations range from those of how people set and respond to personal distances as well as the layout of houses and towns as these are dictated by cultural considerations. See the Glossary of this book and Hall, 1963, 1964, and 1974.

3. There is now a Society for the Anthropology of Visual Communication which constitutes a subfield of anthropology.

4. This research was funded by a grant from the National Institute for Mental Health.

5. John Collier, *Alaskan Eskimo Education*, 1973, and *Visual Anthropology*, 1967.

6. Because of the way in which my research was organized, it was not possible to replicate the university conditions that were more controlled. The test results were close enough, however, to satisfy my needs at the time. Also, the results were consistent with everything else I knew about the two groups.

7. "CBS Reports," January 29, 1981; a one-hour special on Japanese automobile manufacturers.

8. George Leonard, *The Silent Pulse*, 1981.

9. George Leonard, op. cit., 1981.

10. William S. Condon, "Neonatal Entrainment and Enculturation," 1979; William S. Condon and L. W. Sander, "Synchrony Demonstrated Between Movements of the Neonate and Adult Speech," 1974.

11. R. Murray Schafer, *The Tuning of the World*, 1977.

12. Barbara Tedlock, "Songs of the Zuñi Katchina Society: Composition, Rehearsal, and Performance," 1980.

13. Norbert Wiener, *Cybernetics*, 1948.

Chapter 10

1. While synchrony and entrainment appear to mean the same thing, they focus on different aspects of the same process. Synchrony is the manifest observable phenomena; entrainment refers to the internal processes that make this possible, i.e., the two nervous systems "drive each other."

2. Joseph McDowell (1978) made an abortive attempt to replicate Condon's work. If he had done what Condon did and used the same equipment, this might have been possible, but he didn't. For a more complete statement concerning why McDowell failed either to test Condon's assumptions or to replicate his research, see J. B. Gatewood and B. Rosenwein, 1981.

3. William S. Condon, "An Analysis of Behavioral Organization," in *Sign Language Studies*, 13, 1978.

4. Aberrations in self-synchrony can be seen in stuttering, strokes, Parkinson's disease, many forms of "clumsiness," or when people move in an awkward manner.

5. Gay G. Luce and Julius Segal, *Sleep*, 1966.

6. In addition to the example given, Condon, using an oscilloscope, has studied 242 utterances, 365 consecutive words, and 1,055 consecutive phone types with the same results.

7. William S. Condon, "Method of Micro-Analysis of Sound Films of Behavior," 1970; William S. Condon and W. D. Ogston, "Speech and Body Motion Synchrony of Speaker-Hearer," 1971.

8. Edward T. Hall, *The Hidden Dimension*, 1966.

9. William S. Condon, personal communication, 1979.

10. William S. Condon, personal communication.

11. The amount of time depends on the situation and the context. As yet, no one has spelled out the rules. If Mother says she's going to spank Johnny if he comes in one more time with muddy feet, and fails to do so within the half hour of the time she discovers mud in the living room, the chances are she won't spank him at all. If the school principal says that he is going to give a prize of one hundred dollars to the most creative science project and hasn't done so thirty days after the science fair, the chances are he doesn't intend to give the prize. As a general rule, the more that is involved, the more time is allowed before you must act.

12. William S. Condon, "Neonatal Entrainment and Enculturation," 1979.

13. Edward T. Hall, *The Silent Language*, 1959.

14. Carl Gustav Jung, "Synchronicity: An Acausal Connecting Principle," in *The Structure and Dynamics of the Psyche*, second edition, 1969; *Memories, Dreams, Reflections*, revised edition, 1973.

15. Carl Gustav Jung, op. cit., 1973, p. 300.

16. Eliot Chapple, *Culture and Biological Man*, 1970.

Chapter 11

1. *Time*, May 4, 1981, "The Money Chasers."

2. Margaret Mead, *New Lives for Old: Cultural Transformation— Manus, 1928–1953*, 1956.

3. I. I. Rabi, "Introduction," *Time*, 1966.

4. Carlos Fuentes, Honnold Lecture, Knox College, Galesburg, Ill., 1981.

5. Alexander Marschack, *The Roots of Civilization*, 1972.

6. Gerald S. Hawkins, *Stonehenge Decoded*, 1965.

7. E. E. Evans-Pritchard, *The Nuer*, 1940; Paul Bohannan, "Concepts of Time Among the Tiv of Nigeria," 1953.

8. Edward T. Hall, *The Silent Language*, 1959; Benjamin Lee Whorf, *Language, Thought, and Reality*, 1956.

9. William S. Condon's magnum opus is still in preparation. The reader is referred instead to the bibliography of this volume as a beginning.

10. John Collier, *Alaskan Eskimo Education*, 1973; *Visual Anthropology*, 1967.

11. As superintendent of the Sioux Indians, Reifel issued strict instructions that no school bus or any other bus on the reservation was to wait for any Indian. School clocks were repaired and synchronized and schools were run on a strict schedule. Reifel knew that it was better for his tribesmen to miss the bus on the reservation than to miss the job in the white man's town.

12. Richard Rodriguez, *Hunger of Memory*, 1982.

Appendix II

1. The Cooper-Hewitt exhibition devoted to MA (Japanese time-space) was a variation of Arata Isozaki and his collaborator's Paris exhibition, timed to coincide with the "Japan Today" celebration in the late 1970s.

2. General Alexander Haig, President Ronald Reagan's former Secretary of State, caused considerable flurry of comment in the media in October and November 1981 because of personality traits that made him appear ultrasensitive and difficult to deal with. It was at this time that the term "window of vulnerability" was used.

GLOSSARY

Action Chain. A term borrowed from the field of animal behavior to describe an interactional process in which one action releases another in a uniform patterned way. Courtship is a rather complex example. Making a date or inviting someone to dinner would be another. The point is that the two parties play different roles which are interdependent: A. invites B., who must then respond until the paradigm is played out. If the chain is broken at any point, it must begin all over again. Life is full of action chains. In fact, they have not been cataloged, nor has a preliminary list been made even for a single culture. The crucial part as far as humans are concerned is that the steps and stages are unique for each culture (Hall, 1976).

Adumbration. The term adumbration is borrowed from literature and means to foreshadow. Adumbrative processes are related to action chains, except that there is an escalation of intensity and specificity of the message as well as the fact that in an adumbrative situation the individual (or group) will stop the escalation process and select one of a number of action chains which are in the inventory of the culture. Diplomacy in its classical form is the science of reading adumbrative behavior. By the time the public is involved in the process the adumbrative paradigm has progressed a long way—sometimes as far as war. Successful adumbrative practitioners are skillful at getting their message across before "face" and ego become involved. Adumbrations begin at the very high context end of the scale and drop in context (become more explicit) with each step. Adumbrations, like action chains, are culture specific.

AE Peoples. American-European. Current usage to distinguish a cluster of cultural and other traits of AE peoples from Africa, Asia, the subcontinent of India, and native peoples indigenous to North and South America.

Basic Culture. See *Primary Level Culture.*

CII Peoples. Colonial-Iberian-Indian. People originally from Spain and Portugal who migrated to the New World and in many cases mixed with the indigenous populations.

Context, High and *Low.* High context or low context refers to the amount of information that is in a given communication as a function of the context in which it occurs. A highly contexted communication is one in which most of the meaning is in the context while very little is in the transmitted message. A low context communication is similar to interacting with a computer—if the information is not explicitly stated, and the program followed religiously, the meaning is distorted. In the Western world, the law is low context, in comparison with daily transactions of an informal nature. People who know each other over a long period of years will tend to use high context communication.

Enculturation. The process of learning a culture is called enculturation. The enculturation process usually progresses in stages; six-year-olds are more enculturated than three-year-olds, teenagers have almost completed the process and in many cases are under the impression that they have, which can be a source of tension between them and fully enculturated individuals. There are times when the term is confused with *acculturation,* which is the process involving an entire group such as Native Americans, some of whom are so acculturated that it is impossible to distinguish them from any other members of the dominant society.

Entrainment. Entrainment can be observed in the physical as well as the organic world. Fireflies have a tendency toward entrainment, which can be observed as they blink in unison. Electronic oscillators will, if their frequencies are close enough, entrain with the fastest frequency, while pendulum clocks running side by side will entrain if the pendula are the same length. Most people are familiar with the high school experiment in which two tuning forks of the same size will drive each other—that's entrainment. William Condon settled on this term to describe a process that makes syncing possible, wherein one central nervous system drives another, or two central nervous systems drive each other. (See chapter 10.)

Extension Transference. ET is a process whereby an activity or product that is the result of the externalizing—extension—process (Hall, 1959) is confused with the basic or underlying process that has been extended. A classic example is the written form of the language which is commonly treated as the language (Hall, 1976). That is, the map is not the terrain.

Kinesics. The study of body motion as a communicative process (either conscious or unconscious but frequently unconscious). Raymond Birdwhistell originated the terms to distinguish his field from the study of simple gestures. Kinesics is the technical, and correct, term for body language.

Monochronic and *Polychronic Time.* M-time and P-time designate two mutually exclusive kinds of time. M-time is one-thing-at-a-time, following a linear form so familiar in the West. P-time is polychronic, that is, many-things-at-a-time. Schedules are handled quite differently; in fact, there are times when it is difficult to determine whether a schedule exists or not. P-time is common in Mediterranean and CH cultures.

Open and *Closed Score.* See Halprin, 1970. A score is a paradigm— a plan or a set of rules or procedures for accomplishing a task. A simple shopping list is a score, as is a computer program, or an agenda. A closed score is like a computer program—very tightly planned. You succeed if you achieve your objective in the manner specified in advance. Most research is closed score. An open score is just the opposite—you fail if you achieve what you set out to do in the way in which you originally planned. Open scores are spontaneous. The most creative research and practically all scientific breakthroughs are open score, at least in the initial phases. If, when using an open score, nothing new has been introduced, one fails. Closed scores are carefully programmed; open scores are spontaneous, intuitive, and innovative (Hall, 1976).

Primary Level Culture. There are at least three readily identifiable levels of culture: *primary, secondary,* and *explicit* or *manifest.* Basic primary level culture—BPL culture—is that variety of culture in which the rules are: known to all, obeyed by all, but seldom if ever stated. Its rules are implicit, taken for granted, almost impossible for the average person to state as a system, and generally out of awareness. *Secondary level culture,* though in full awareness, is normally hidden from outsiders. Secondary level culture is as regular and binding as any other level of culture, possibly even more so. It is that level of culture which the Pueblo Indians of New Mexico keep from white people. But it can also be the special culture of virtually any group or society. *Tertiary* or *explicit, manifest culture* is what we all see and share in each other. It is the façade presented to the world at large. Because it is so easily manipulated, it is the least stable and least dependable for purposes of decision making. Most social science and political science these days is directed at strategies for penetrating the screen separating manifest culture from secondary level culture.

Proxemics. Proxemics is the study of people's use of space as a function of culture. That is, the effect of culture on the structuring and use of space. Personal distancing and the unstated rules for laying out houses and towns. (See Hall, 1966.)

Releaser. The concept of releasers is intimately tied up with communication theory as it applies to culture. Implicit in the work of Charles Hockett (1958 and 1964), the releaser theory fits well into the high and low context paradigm. As the context lowers and communication becomes increasingly technical—and lengthy—the releasers become more complex. While many releasers are linguistic in nature, they are not restricted to language. In fact, they make the whole process of communication conform to what is known of culture so that it is possible to consider language and culture as components of the same process. The point about the releaser theory of communication is that the releaser is just that: language releases a response that is already programmed (either genetically or through learning) in the other party. When it comes to communication, there is no such thing as a tabula rasa as far as the human species is concerned.

Right Brain, Left Brain. A concept popularized by Robert Ornstein and others concerning a specialized, differentiated function of the right and left hemispheres of the cerebral cortex. In general, in the Western world, the left hemisphere is word- and numbers-oriented, and more linear than the right hemisphere, which is holistic and spatial in its organization. There is some indication that the Japanese do not divide the functions of the two hemispheres in the same way that we do in the West.

Society or Social Organization. Used technically in this sense to designate the institutions or the manner in which they are organized, or structured, as contrasted with the culture. The distinction may be an artifact of our own culture in America since the British anthropologists use society in a way that is similar to the way Americans refer to culture. I think of American society or the American social system as comparable to an organization chart, whereas culture is more closely related to what the people in that organization do when they perform their functions. The distinction may in the long run turn out to be a convention or merely a convenience.

Surface, Explicit, or *Manifest Culture.* See *Primary Level Culture.*

Syncing or *In Sync.* The term "in sync" came out of the need to synchronize the sound track of a cinema with the pictures—hence the term "sync sound." In recent years, due to the research of men like Condon and Birdwhistell which demonstrates that human beings synchronize with each other just as precisely as the sound technician synchronizes his sound track with a film, this feature of human behavior has been referred to as syncing, or to be in sync.

BIBLIOGRAPHY

Abernathy, William J., and Hayes, Robert. "Managing Our Way to Economic Decline." *Harvard Business Review*, July 1980.

Aschoff, Jurgen. "Circadian Rhythms in Man." *Science*, Vol. 148, June 11, 1965.

Ayensu, Edward S., and Whitfield, Philip. *The Rhythms of Life*. New York: Crown Publishers, 1982.

Barnett, Lincoln. *The Universe and Dr. Einstein*. New York: William Sloane Associates, 1950.

Benedict, Ruth. *The Chrysanthemum and the Sword*. Boston: Houghton Mifflin Co., 1946.

———. *Zuñi Mythology*, 2 vols. New York: AMS Press, 1969.

Birdwhistell, Raymond. *Introduction to Kinesics*. Louisville, Ky.: University of Louisville Press, 1952, 1974.

Bohannan, Paul. "Concepts of Time Among the Tiv of Nigeria." *Southwestern Journal of Anthropology*, Vol. 9, No. 3, Autumn 1953.

Boyd, Doug. *Rolling Thunder*. New York: Dell Publishing Co., 1974.

Brazelton, Thomas Berry. *On Becoming a Family: The Growth of Attachment*. New York: Delacorte Press, 1981.

Brodey, Warren. "The Clock Manifesto." In Roland Fischer, ed., *Inter-*

disciplinary Papers of Time. Proceedings of the New York Academy of Science, Annual Meeting, 1966. New York: New York Academy of Science, 1967.

Brown, Frank A. "Living Clocks." *Science*, Vol. 130, December 4, 1959.

Bruneau, Thomas J. "The Time Dimension in Intercultural Communication." *Communication Yearbook 3*, Dan Nimo, ed. New Brunswick, N.J.: Transaction Books, 1979.

Capra, Fritjof. *The Tao of Physics.* New York: Bantam Books, 1977.

Carrington, Patricia. *Freedom in Meditation.* Garden City, N.Y.: Anchor Books/Doubleday, 1977.

Chapple, Eliot. *Culture and Biological Man.* New York: Holt, Rinehart & Winston, 1970.

Church, Margaret. *Time and Reality: Studies in Contemporary Fiction.* Chapel Hill, N.C.: University of North Carolina Press, 1963.

Collier, John. *Alaskan Eskimo Education: A Film Analysis of Cultural Confrontation in the Schools.* New York: Holt, Rinehart & Winston, 1973.

————. *Visual Anthropology: Photography as a Research Method.* New York: Holt, Rinehart & Winston, 1967.

Colton, Harold S. *Hopi Kachina Dolls.* Albuquerque, N.M.: University of New Mexico Press, 1949.

Condon, William S. "An Analysis of Behavioral Organization." *Sign Language Studies*, 13, 1978.

————. "Method of Micro-Analysis of Sound Films of Behavior." *Behavior Research Methods and Instrumentation*, Vol. 2, 1970.

————. "Multiple Response to Sound in Dysfunctional Children." *Journal of Autism and Childhood Schizophrenia*, Vol. 5, 1975.

————. "Neonatal Entrainment and Enculturation." In M. Bullowa, ed., *Before Speech: The Beginning of Interpersonal Communication*, New York: Cambridge University Press, 1979.

————. "A Primary Phase in the Organization of Infant Responding Behavior." In H. R. Schaffer, ed., *Studies in Mother-Infant Interaction*, New York: Academic Press, 1977.

————, and Brosin, H. W. "Micro Linguistic–Kinesic Events in Schizophrenic Behavior." In D. V. S. Sankar, ed., *Schizophrenia: Current Concepts and Research*, Hicksville, N.Y.: PJD Publications, 1969.

————, and Ogston, W. D. "A Method of Studying Animal Behavior." *Journal of Auditory Research*, Vol. 7, 1967b.

————, and Ogston, W. D. "A Segmentation of Behavior." *Journal of Psychiatric Research*, Vol. 5, 1967a.

————, and Ogston, W. D. "Sound Film Analysis of Normal and Pathological Behavior Patterns." *Journal of Nervous and Mental Disease*, Vol. 143, 1966.

————, and Ogston, W. D. "Speech and Body Motion Synchrony of Speaker-Hearer." In D. L. Horton and J. J. Jenkins, eds., *Perception of Language*, Columbus, Ohio: Charles E. Merrill Press, 1971.

————, and Sander, L. W. "Neonate Movement Is Synchronized with Adult Speech: Interactional Participation and Language Acquisition." *Science*, Vol. 183, 1974a.

————, and Sander, L. W. "Synchrony Demonstrated Between Movements of the Neonate and Adult Speech." *Child Development*, Vol. 45, 1974b.

Conklin, J. C. *Folk Classification*. New Haven, Conn.: Yale University Press, 1972.

————, and Saito, Mitsuko. *Intercultural Encounters with Japan*. Tokyo: Simul Press, 1974.

Danielli, Mary. "The Anthropology of the Mandala." *The Quarterly Bulletin of Theoretical Biology*, Vol. 7, No. 2, 1974.

Dean, Terrence, and Kennedy, Allan. *Corporate Cultures*. Reading, Mass.: Addison-Wesley Publishing Co., 1982.

De Grazia, S. *Of Time, Work and Leisure*. New York: Twentieth Century Fund, 1962.

De Long, Alton. "Phenomenological Space-Time: Toward an Experimental Relativity." *Science*, 213, August 7, 1981.

————. "Spatial Scale and Perceived Time-Frames: Preliminary Notes on Space-Time in Behavioral and Conceptual Systems." Knoxville, Tenn., no date.

————. "The Use of Scale Models in Spatial-Behavioral Research." *Man-Environment Systems*, Vol. 6, 1976.

————, and Lubar, J. F. *Scale and Neurological Function, Summary*. Knoxville, Tenn.: University of Tennessee Press, 1978.

Dewey, John. *Art as Experience*. New York: G. P. Putnam's Sons, 1934, 1959.

Doob, Leonard. "Time: Cultural and Social Aspects." In T. Carlstein, D. Parkes, and N. Thrift, eds., *Making Sense of Time*, Vol. 1, London: Edward Arnold, 1978.

Dürckheim, Karlfried Graf von. *Hara: The Vital Centre of Man*. London: George Allen & Unwin, 1962.

Einstein, Albert. *Relativity: The Special and General Theory*. Trans. by Robert W. Lawson. New York: Henry Holt & Company, 1920.

Ekman, Paul. *Emotion in the Human Face*. New York: Pergamon Press, 1972.

218   BIBLIOGRAPHY

Eliade, Mircea. *The Sacred and the Profane*. New York: Harcourt, Brace and World, 1959.

Englebardt, Stanley L. "The Marvels of Microsurgery." *The Atlantic*, February 1980.

Evans-Pritchard, E. E. *The Nuer*. Oxford, England: Clarendon Press, 1940.

Fallows, James. "American Industry. What Ails It, How to Save It." *The Atlantic*, September 1980.

Floyd, Keith. "Of Time and Mind: From Paradox to Paradigm." In John White, ed., *Frontiers of Consciousness*, New York: Avon Books, 1974.

Fraser, Julius T., ed. *The Voices of Time*. New York: George Braziller, 1966. Amherst, Mass.: University of Massachusetts Press, 2nd ed., 1981.

Frazier, K. "The Anasazi Sun Dagger." *Science 80*, November/December 1979.

Fromm, Erich. *Man for Himself*. New York: Rinehart & Co., 1947.

———, Suzuki, D. T., and others. *Zen Buddhism and Psychoanalysis*. New York: Harper & Brothers, 1960.

Fuentes, Carlos. Honnold Lecture, given at Knox College, Galesburg, Ill. Knox Alumnus 7, October 15, 1981.

Gardner, Howard. "Thinking: Composing Symphonies and Dinner Parties." *Psychology Today*, Vol. 13, No. 1, April 1980.

Gatewood, J. B., and Rosenwein, R. "Interactional Synchrony: Genuine or Spurious? A Critique of Recent Research." *Journal of Nonverbal Behavior*, Vol. 6, No. 1, 1981.

Gedda, Luigi, and Brenci, Gianni. *Chronogenetics*. Louis Keith, ed. Springfield, Ill.: Charles C. Thomas Publisher, 1978.

Gibney, Frank. *Japan, the Fragile Superpower*. New York: New American Library, 1979.

Hall, Edward T. "Adumbration as a Feature of Intercultural Communication," *American Anthropologist*, Vol. 66, No. 6, December 1964.

———. *Beyond Culture*. Garden City, N.Y.: Anchor Press/Doubleday, 1976.

———. *Handbook for Proxemic Research*. Washington, D.C.: Society for the Anthropology of Visual Communication, 1974.

———. *The Hidden Dimension*. Garden City, N.Y.: Doubleday & Company, 1966.

———. *The Silent Language*. Garden City, N.Y.: Doubleday & Company, 1959.

————. "A System for the Notation of Proxemic Behavior." *The American Anthropologist*, Vol. 65, No. 5, October 1963.

Halprin, Lawrence. *The R.S.V.P. Cycles; Creative Processes in the Human Environment*. New York: George Braziller, 1970.

Hardin, Garrett. *Exploring New Ethics for Survival: The Voyage of the Spaceship "Beagle."* New York: The Viking Press, 1972.

Hawkins, Gerald S., in collaboration with John B. White. *Stonehenge Decoded*. Garden City, N.Y.: Doubleday & Company, 1965.

Hediger, Heini. *Studies of the Psychology and Behaviour of Captive Animals in Zoos and Circuses*. Trans. by Geoffrey Sircom. London: Butterworth & Co., 1955.

Hennig, Margaret, and Jardim, Anne. *The Managerial Woman*. Garden City, N.Y.: Anchor Press/Doubleday, 1977.

Herrigel, Eugen. *Zen in the Art of Archery*. Trans. by R. F. C. Hull. New York: Vintage Books, 1971.

Hoagland, Hudson. "Brain Evolution and the Biology of Belief." *Science*, Vol. 33, No. 3, March 1977.

Hockett, Charles F. *A Course in Modern Linguistics*. New York: The Macmillan Company, 1958.

————, and Asher, R. "The Human Revolution," *Current Anthropology*, Vol. 5, No. 3, 1964.

Hoffmann, Yoel. *Every End Exposed: The 100 Koans of Master Kidō*. Brookline, Mass.: Autumn Press, 1977.

Howard, N. E. *Territory and Bird Life*. London: John Murray, 1920.

Isozaki, Arata. *MA: Space-Time in Japan*. Catalog for the 1979 Cooper-Hewitt Museum Exhibition on MA. New York.

James, H. C. *Pages from Hopi History*. Tucson, Ariz.: University of Arizona Press, 1974.

Jung, Carl Gustav. *Memories, Dreams, Reflections*. New York: Pantheon Books, revised edition, 1973.

————. "Synchronicity: An Acausal Connecting Principle," in *The Structure and Dynamics of the Psyche*, Bollingen Series XX, Vol. 8, Princeton, N.J.: Princeton University Press, second edition, 1969.

————, and Pauli, Wolfgang. *The Interpretation of Nature and the Psyche*. New York: Pantheon Books, 1955.

Kabotie, Fred. *Fred Kabotie, Hopi Indian Artist*. Flagstaff, Ariz.: Museum of Northern Arizona, 1977.

Kardner, Abraham. *The Psychological Frontiers of Society*. New York: Columbia University Press, 1945.

Kasamatsu, Akira, and Hirai, Tomio. "An Electroencephalographic

Study on the Zen Meditation (Zazen) Folio." *Psychiatia Neurologia Japonica*, Vol. 20, 1966. (Seishin Shinkeigaku Zasshi)

Kato, Hidetoshi. "Mutual Images: Japan and the United States Look at Each Other." In Condon and Saito, eds., *Intercultural Encounters with Japan*, Tokyo: Simul Press, 1974.

Kennard, Edward H. *Hopi Kachinas*. New York: J. J. Augustin, 1938.

Kierkegaard, Søren. *The Concept of Dread*. Trans. by Walter Lowrie. Princeton, N.J.: Princeton University Press, 1944.

Kilpatrick, Franklin P., ed. *Explorations in Transactional Psychology*. New York: New York University Press, 1961.

Korda, Michael. *Success!* New York: Random House, 1977.

————. *Power: How to Get It, How to Use It*. New York: Random House, 1975; Ballantine Books, 1976.

Korzybski, Count Alfred. *Science and Sanity: An Introduction to Non-Aristotelian Systems and General Semantics*. Lakeville, Conn.: International Non-Aristotelian Library Publishing Company, 1948.

Leach, E. R. *Rethinking Anthropology*. London: Athlone Press, 1961.

Le Lionnais, François. "Le Temps." In Robert Delpire, ed., Encyclopédie Essentielle, Paris, 1959.

Leonard, George. *The Silent Pulse*. New York: Bantam Books, 1981.

Libassi, Paul T. "Biorhythms: The Big Beat." *The Sciences*, May 1974.

Lorenz, Konrad. *King Solomon's Ring*. New York: The Thomas Y. Crowell Co., 1952.

————. *Man Meets Dog*. Cambridge, Mass.: Riverside Press, 1955.

Luce, Gay G. *Body Time: Physiological Rhythms and Social Stress*. New York: Random House, 1971.

————, and Segal, Julius. *Sleep*. New York: Coward-McCann, 1966.

Mabie, H. W. *In the Forest of Arden*. New York: Dodd, Mead & Co., 1898.

MacLean, Paul D. "Man and His Animal Brains." *Modern Medicine*, Vol. 95, 1965, p. 106.

Maraini, Fosco. *Japan: Patterns of Continuity*. Tokyo: Kodansha International, 1979.

Marschack, Alexander. "Ice Age Art." *Explorers Journal*, Vol. 59, No. 2, June 1981.

————. *The Roots of Civilization*. New York: McGraw-Hill Book Co., 1972.

Matsumoto, M. "Haragei" (ms.). 1981.

Mayer, Maurice. *The Clockwork Universe: German Clocks and Automata, 1550–1650*. Neal Watson Academic Publications, New York, 1980.

BIBLIOGRAPHY 221

McDowell, Joseph J. "Interactional Synchrony: A Reappraisal." *Journal of Personal and Social Psychology*, Vol. 35, No. 9, 1978.

Mead, Margaret. *New Lives for Old: Cultural Transformation—Manus, 1928–1953*. New York: William Morrow & Co., 1956.

Munro, H. H. "The Unrest Cure," The Chronicles of Clovis. In *The Short Stories of Saki*, New York: The Viking Press, 1946.

Needham, J. "Time and Eastern Man." Royal Anthropological Institute, Occasional Paper, No. 2, 1965.

Nystrom, Christine. "Mass Media: The Hidden Curriculum." Educational Leadership, November 1975.

Ornstein, Robert. *On the Experience of Time*. Baltimore, Md.: Penguin Books, 1969.

———. *The Psychology of Consciousness*. New York: Pelican Books, 1975.

Park, David. *The Image of Eternity*. Amherst, Mass.: University of Massachusetts Press, 1975.

Piaget, Jean. *The Child's Conception of Time*. Trans. by A. J. Pomerans. New York: Basic Books, 1970.

———. *The Child's Conception of the World*. Trans. by Joan and Andrew Tomlinson. New York: Harcourt, Brace & Co., 1929.

———. "Time Perception in Children." In Julius T. Fraser, *The Voices of Time*, 1981, trans. by Betty Montgomery, ed. by Emily Kirb.

———, and Inhelder, Bärbel. *The Child's Conception of Space*. Trans. by F. J. Langdon and J. L. Lunzer. London: Routledge & Kegan Paul, 1956.

Piers, M. W. "Editorial." Erikson Institute Outrider, No. 18, Chicago, Fall 1980.

Pietsch, Paul. *Shufflebrain*. Boston: Houghton Mifflin Co., 1981.

Powers, William T. *Behavior: The Control of Perception*. Chicago: Aldine Publishing Co., 1973.

———. "Beyond Behaviorism." *Science*, Vol. 179, January 26, 1973.

Pribram, Karl H. *Languages of the Brain*. Englewood Cliffs, N.J.: Prentice-Hall, 1971.

Priestley, J. B. *Man and Time*. Garden City, N.Y.: Doubleday & Company, 1964.

Puthoff, H. E., and Targ, R. "Perceptual Channel for Information Transfer Over Kilometer Distances: Historical Perspective and Recent Research." Proceedings of the IEEE, Vol. 64, No. 3, March 1976, pp. 329–54.

Rabi, I. I. "Introduction," *Time*. New York: Time-Life Books, 1966.

Rodriguez, Richard. *Hunger of Memory*. Boston: David R. Godine, 1982.

Scarf, Maggie. *Unfinished Business: Pressure Points in the Lives of Women*. Garden City, N.Y.: Doubleday & Company, 1980.

Schafer, R. Murray. *The Tuning of the World*. New York: Alfred A. Knopf, 1977.

Scheflen, Albert E. *Body Language and the Social Order*. Englewood Cliffs, N.J.: Prentice-Hall, 1972.

Searles, Harold. *The Non-Human Environment*. New York: International Universities Press, 1956.

Selye, Hans. *The Stress of Life*. New York: McGraw-Hill Book Co., 1956.

Shaw, George Bernard. *Cashel Byron's Profession*. New York: Harper & Brothers, 1886.

Sheldon, William, and Gruen, S. S. *Varieties of Human Temperament: A Psychology of Constitutional Differences*. New York: Hafner Publishing Co., 1970.

Simmons, Leo W., ed. *Sun Chief: The Autobiography of a Hopi Indian*. New Haven, Conn.: Yale University Press, 1942.

Skinner, B. F. "Selection by Consequences." *Science*, Vol. 213, July 31, 1981.

Slovenko, Ralph. "Public Enemy No. 1 to Community and Mental Health: The Automobile." Bulletin of the American Academy of Psychiatry and Law, Vol. 4, 1976, p. 287.

Sofaer, A., Sinclair, R. M., and Doggett, L. E. "Lunar Markings on Fajada Butte in Chaco Canyon, New Mexico." In *New World Archaeoastronomy*, Cambridge, England: Cambridge University Press, 1982.

————, Zinser, V., and Sinclair, R. M. "A Unique Solar Marking Construct." *Science*, October 19, 1979.

Sol, D. R. "Timers in Developing Systems." *Science*, Vol. 203, March 1979.

Suzuki, D. T. *Essays on Zen Buddhism*. London: Rider & Co., 1951.

————. *Manual of Zen Buddhism*. London: Rider & Co., 1950.

————. *Zen and Japanese Culture*. Princeton, N.J.: Bolingen Foundation, Princeton University Press, 1959.

————. *Zen Buddhism*. Garden City, N.Y.: Doubleday & Company, 1956.

————, and Fromm, Erich. *Zen Buddhism and Psychoanalysis*. New York: Harper & Brothers, 1960.

Tedlock, Barbara. "Songs of the Zuñi Katchina Society: Composition,

Rehearsal, and Performance." In Charlotte Frisbie, ed., *Southwestern Indian Ritual Drama*, Albuquerque, N.M.: University of New Mexico Press, 1980.

———. *Time and the Highland Maya*. Albuquerque, N.M.: University of New Mexico Press, 1981.

Tsunoda, Tadanobu. *Nihon-jin No No—The Japanese Brain*. Tokyo: 1978.

UNESCO. *Cultures and Time: At the Cross Roads of Culture*. Paris: UNESCO Press, 1976.

Vogel, Ezra. *Japan as Number 1*. New York: Harper & Row, 1979.

Von Uexkull, Jacob. "A Stroll Through the Worlds of Animals and Men." In C. Schiller, ed., *Instinctive Behavior*, New York: International Universities Press, 1964.

———, and Kriszat, Georg. *Streifzüge durch die Umwelten von Tieren und Menschen*. Berlin: J. Springer, 1934.

Watts, Alan. *The Way of Zen*. New York: Pantheon Books, 1957.

Whorf, Benjamin Lee. *Language, Thought, and Reality*. New York: John Wiley & Sons, 1956.

———. "Science and Linguistics." *The Technology Review*, Vol. XLII, No. 6, April 1940.

Wiener, Norbert. *Cybernetics: or, Control and Communication in the Animal and the Machine*. New York: John Wiley & Sons, 1948.

Yamaoka, Haruo. *Meditation But Enlightenment: The Way of Hara*. Tokyo: Heian International Publishing Co., 1976.

Zerubavel, Eviatar. *Hidden Rhythms*. Chicago: University of Chicago Press, 1981.

INDEX

Action chain: defined, 209
Adumbration, 158, 169; defined, 209–210
American-European (AE) culture (The West), 24, 42, 49, 100–13, 143, 176–86, 194–95, 200 n.; artists, contrasted to Japanese, 92–94; business practices, 61–62, 71, 102–104, 107–8, 150; calendar contrasted with Quiché, 77–79; closure, 31; concept of time, contrasted to Quiché, 76–80, 81–82; contrasted to Japanese culture, 85–99; conversational distance, 140–141; defined, 210; drive to proselytize, 80; duality, 124; fast and slow messages, 60–61; feedback rhythm compared to Spanish, 159–60; languages and time, 35; law, 63, 169, 179; marriage, 37; preoccupation with variable time, 119–23; rhythm system compared to Native American, 169–71; routines, 75; symbolic meaning of time, 68–69; thinking contrasted to Zen Buddhist, 91–92; time compression and time expansion, 125–26; view of archery, contrasted to Japanese, 90–91; view of past and future, 130, 183, 201 n.; waiting, 122. *See also* French, German, Japanese cultures, Monochronic time, Women

American Indians (Native Americans), 25, 41, 42, 80–81, 142, 157, 167–68, 171, 172. *See also* Hopi, Navajo, Pueblo Indians, Santo Domingo Pueblo, Shoshone, Taos Pueblo, Zuñi Pueblo

Arab cultures, 42, 44, 46, 123. *See also* Polychronic time

Archery, 90–91

Art, 92–94. *See also* American-European culture, Zen Buddhism

Astrology, 78

Athletes, 127, 139, 152–53, 205 n.

Bateson, Gregory, 163
Beethoven, Ludwig van, 63, 128–29, 199 n.
Benedict, Ruth: *The Chrysanthemum and the Sword*, 85, 197 n., 201 n., 202 n.; *Zuñi Mythology*, 20, 197 n.

Latin American culture, 44, 46. *See also* Polychronic time

Leach, E. R.: English anthropologist, 5, 197 n.

Leonard, George: on rhythm, 150–151, 153, 166, 206 n.

Libassi, Paul T.: "Biorhythms: The Big Beat," 197 n.

Lieberson, Goddard: "They Said It With Music," 157

Linear logic, 8, 132. *See also* Piaget

Luce, Gay G.: *Sleep*, 206 n.

Ma: Japanese concept, 86, 92, 189–195, 208 n.; defined and characterized, 189–90; *hashi*, 191; *himorogi*, 190–91; *michiyuki*, 193–94; *sabi*, 193; *suki*, 192; *susabi*, 193; *utsuroi*, 192–93; *utsushimi*, 193; *yami*, 191. *See also* Zen Buddhism

McDowell, Joseph, 206 n.

MacLean, Paul D.: on the brain, 199 n.

McQueen, Steve: *On Any Sunday*, 152

Mandala, 15–16, 187–88. *See also* Time

Mann, Judy: column in Washington *Post*, 199 n.

Maori, 173

Maraini, Fosco: *Japan: Patterns of Continuity*, 201 n.

Marschack, Alexander: archaeologist of Stone Age, 118, 184, 203 n., 207 n.

Marxism, 8

Matsuda, Takeo: Japanese housing developer, 64–65

Matsumoto, Michihiro: Japanese author, 95, 201 n., 202 n.

Mayan calendar, 76–77

Mead, Margaret: *People of Manus*, 182, 207 n.

Meaning, 56–57. *See also* Context

Meditation, 19, 135–36

Message velocity spectrum, 59–61, 63, 199 n.

Michi: Japanese concept, 92. *See also* Japanese culture

Microsurgery, 127

Miller, Arthur: *Death of a Salesman*, 176–77

Monochronic time, 24, 41–53, 71, 104, 109–13, 147, 185, 200 n.; American-European pattern, 44–45; defined, 43, 45, 211; example of Japanese use, 51–54; German pattern, 106–7; New Mexico Spanish and, 65–66; strengths and weaknesses, 48; task orientation, 50, 70–72; television commercial as example, 47–48. *See also* American-European culture, Bureaucracy

Mozart, Wolfgang Amadeus, 128–29

Munro, H. H. (Saki), 75

Music, 63, 89, 128–29, 155–57, 162, 172–73. *See also* Rhythm, Synchrony

Navajo, 28, 92, 157, 198 n.; concept of time, 28; different from Hopi in relationship to work, 32–33; waiting, 122. *See also* American Indians, Time

Neanderthal, 117, 203 n.

Nemawashi: Japanese concept, 89. *See also* Japanese culture

New Mexico Spanish culture, 42, 65–66, 142, 147, 158; feedback rhythm, 159–60. *See also* Polychronic time

Newton, Isaac, 13, 20–21, 22, 132

Newtonian model of time, 5, 21, 123, 201 n.

Nine to Five: film with Lily Tomlin, Jane Fonda, and Dolly Parton, 157

Nonverbal behavior (communication), 144–48. *See also* Film Research, Proxemics, Synchrony

Northwestern University, 145

Nuer (African people), 5, 74; structure of time, 74. *See also* Time

Ornstein, Robert, 8, 197 n., 201 n.

Palmer, Gabrielle, 200 n.

Perceptual distortions, 136–37, 166–167